AF789975

Startup Inventor's Handbook

Thomas Heinz Meitinger

Startup Inventor's Handbook

Develop and protect ideas

Springer

Thomas Heinz Meitinger
Meitinger Patentanwalts GmbH
Munich, Germany

ISBN 978-3-662-72136-0 ISBN 978-3-662-72134-6 (eBook)
https://doi.org/10.1007/978-3-662-72134-6

Translation from the German language edition: "Startup Erfinderhandbuch" by Thomas Heinz Meitinger, © Der/die Herausgeber bzw. der/die Autor(en), exklusiv lizenziert an Springer-Verlag GmbH, DE, ein Teil von Springer Nature 2025. Published by Springer Berlin Heidelberg. All Rights Reserved.

This book is a translation of the original German edition "Startup Erfinderhandbuch" by Thomas Heinz Meitinger, published by Springer-Verlag GmbH, DE in 2025. The translation was done with the help of an artificial intelligence machine translation tool. A subsequent human revision was done primarily in terms of content, so that the book will read stylistically differently from a conventional translation. Springer Nature works continuously to further the development of tools for the production of books and on the related technologies to support the authors.

© The Editor(s) (if applicable) and The Author(s), under exclusive license to Springer-Verlag GmbH, DE, part of Springer Nature 2025

This work is subject to copyright. All rights are solely and exclusively licensed by the Publisher, whether the whole or part of the material is concerned, specifically the rights of translation, reprinting, reuse of illustrations, recitation, broadcasting, reproduction on microfilms or in any other physical way, and transmission or information storage and retrieval, electronic adaptation, computer software, or by similar or dissimilar methodology now known or hereafter developed.
The use of general descriptive names, registered names, trademarks, service marks, etc. in this publication does not imply, even in the absence of a specific statement, that such names are exempt from the relevant protective laws and regulations and therefore free for general use.
The publisher, the authors and the editors are safe to assume that the advice and information in this book are believed to be true and accurate at the date of publication. Neither the publisher nor the authors or the editors give a warranty, expressed or implied, with respect to the material contained herein or for any errors or omissions that may have been made. The publisher remains neutral with regard to jurisdictional claims in published maps and institutional affiliations.

This Springer imprint is published by the registered company Springer-Verlag GmbH, DE, part of Springer Nature.
The registered company address is: Heidelberger Platz 3, 14197 Berlin, Germany

If disposing of this product, please recycle the paper.

Preface

A startup is a company that focuses on the implementation of a new technology. The founder of a startup will initially engage with the existing technology, carefully studying its characteristics and, in particular, its disadvantages. In the next step, the founder will define a technical task aimed at overcoming the shortcomings of the current technology. The founder must also conduct a business evaluation to determine whether solving this task will result in a product with sufficient market potential. Only then does the creative process of developing an invention begin.

The development of one's own technical solution can be supported by patent law and the state of the art. The technical solution found should be compared with the state of the art. If third-party patents that are in force are identified, one's own technology must be further developed in order to move outside the scope of protection of those patents. This (forced) further development can be seen as an opportunity to obtain a patentable technical teaching of one's own.

In addition, the founder of a startup should consider specialized embodiments that are suitable for specific applications. These special applications can be derived from the requirements of the state of the art. The developed embodiments can open up additional markets and enhance one's own patent application, as they can be claimed as dependent claims. Engaging with patent law can therefore lead to a sophisticated proprietary technology.

It is advisable for the startup founder to have a trademark protected. The registration of a trademark is not expensive, can be maintained as long as you like, and increases in value as the business grows. This book provides tips on how to develop valuable trademarks. The topic of design rights is also introduced, as design rights are at least relevant as supplementary protective rights.

At the end of the chapters on patent law, trademark law, and design law, you will find specific instructions on how to independently prepare a patent application, a trademark application, or a design application. Chapter 7 summarizes the essence of the book by explaining the development of a proprietary technology through engagement with patent law. This book is application-oriented, but also provides in-depth insight into the theoretical aspects of intellectual property protection, enabling readers to make well-informed decisions.

This book is intended not only to help a startup founder address legal challenges, but also to provide them with the tools necessary to further develop their own technology and, potentially, to obtain a patent as an economic monopoly.

Munich
in November 2024

Dr. Thomas Heinz Meitinger

Contents

About the Author

Dr. Thomas Heinz Meitinger is a German and European patent attorney. He is the managing director of Meitinger Patentanwalts GmbH. Meitinger Patentanwalts GmbH is a medium-sized patent law firm based in Munich. After studying electrical engineering in Karlsruhe, he initially worked as a development engineer. Subsequent positions included roles as production manager and technical director in medium-sized companies. Dr. Meitinger regularly publishes scientific articles, writes specialist books on intellectual property law, and gives lectures on patent, trademark, and design law. Dr. Meitinger holds the degrees Dipl.-Ing. (Univ.) and Dipl.-Wirtsch.-Ing. (FH). In addition, he holds the following master's degrees: LL.M., LL.M., MBA, MBA, M.A., and M.Sc.

List of Figures

Introduction 1

Contents

This introductory chapter presents the sections of the book along with their content and explains their significance in the process of founding a startup.

A startup is a company whose business focus is the application of a technology as an offering to the market. Accordingly, dealing with the technology and securing legal protection are of particular importance for the startup company.

As the book progresses, the reader will realize that clarifying the legal situation—by examining whether the technology can be used without infringing third-party patents and whether the technology itself is patentable—can lead to further development of the technology, resulting in proprietary patents and increased market opportunities.

The first step is to determine the relevant state of the art. The startup must engage with existing technology and strive to improve it and make it marketable. The founder will then arrive at an initial technical solution.

This initial technical solution must then be examined to determine whether it can be used. It is possible that third-party patents may be found that would be infringed. In such cases, the technology must be further developed to move outside the scope of protection of those patents. It is precisely through this necessary further development that the technology can be improved and lead to a patentable technical teaching.

© The Author(s), under exclusive license to Springer-Verlag GmbH, DE, part of Springer Nature 2025

T. H. Meitinger, *Startup Inventor's Handbook*,
https://doi.org/10.1007/978-3-662-72134-6_1

The final step is to assess patentability. If the technology is eligible for protection, it should be patented in order to secure a unique selling proposition on a lasting basis.

The steps described above demonstrate that it is precisely through the interplay with patent law that a technology can be developed which is suitable for founding a successful startup.

1.1 State of the Art

The state of the art encompasses all documents, publications, and public uses that were made available to the public before the filing or priority date of a patent application.

For a startup, the state of the art is a central point of reference in several respects. On the one hand, the state of the art can help improve the startup's own technology. In particular, by reviewing the state of the art, a problem to be solved may be identified, or the state of the art may reveal a technical solution superior to the startup's own technology. Furthermore, the state of the art can be used to determine whether the technology can be used without infringing third-party intellectual property rights. Finally, the state of the art serves as the basis for assessing whether the technology is patentable.

1.2 Freedom to Operate

With the identified state of the art, it must be examined whether the technology can be used without infringing third-party intellectual property rights. Only those rights that are still in force need to be considered. The challenge here is that one can never be certain that a search is complete. There is always a risk that a relevant document has been overlooked or could not be found, for example, because it was not yet available electronically.

However, it may be possible to find a patent document whose maximum term has expired and which discloses all features of the technology. In this case, it can be assumed that the technology can be used without infringing third-party patents, since the document already constitutes prior art in the public domain that anyone may use.

1.3 Patent Protection

When assessing patentability, particular attention must be paid to the novelty and inventive step of the invention. An invention can only be patentable if it is new and inventive over the state of the art.

1.4 Trademark, Design, and Employee Invention Law

Every company should secure a trademark at the patent office. In particular, a German or EU-wide trademark (Union trademark) may be sought. Without trademark protection, there is a risk that a free rider could register the same trademark and acquire an exclusive right that could be successfully enforced even against the original user of the trademark.

Trademark registration can be obtained at very low cost and can increase in value with each year of business activity, so that a trademark can become a significant and valuable asset for a company. Design rights can at least provide supplementary protection. Employee invention law is relevant if an employee is engaged in inventive activity.

Patent 2

Contents

A technical invention can be protected by a patent or a utility model. Patents and utility models are important for the startup founder, because a startup is technology-centered. Third-party patents can impair the economic activity of the startup company, and own patents can prevent competition from other companies.

© The Author(s), under exclusive license to Springer-Verlag GmbH, DE, part of Springer Nature 2025

T. H. Meitinger, *Startup Inventor's Handbook*,
https://doi.org/10.1007/978-3-662-72134-6_2

2.1 Purpose and Justification of Patent Law

Essentially, the justification for an economic monopoly for an invention is the assumption that the inventor requires legal protection in order to reap the rewards of his invention. If this monopoly were not granted, the inventor would cease his inventive activity. In reality, however, it is rare for individual inventors to file a patent application. In most cases, it is large companies that act as applicants and thereby consolidate their market position.[1] From a macroeconomic perspective, patent law can therefore be viewed quite critically.

Granting an economic monopoly for an invention as a reward and incentive for the inventor follows from natural law, which states that an invention is the property of his inventor. For this reason, the inventor must be granted an exclusive right, allowing them to exclude any unauthorized third party from the use of "his" invention. Accordingly, the Patent Act grants the inventor a claim for injunctive relief.[2]

Patent law and competition law are in part diametrically opposed. Patent law is even fundamentally questioned, as its benefit is said to be lower than the costs incurred and, therefore, patent law would, on balance, harm an economy.[3] Furthermore, economies without patent law or with weak patent law would develop better in particular.[4]

The purpose of patent law from the perspective of the state is to promote the development and dissemination of industrially applicable technical inventions. It is not the task to promote abstract theories. Instead, the public is to be provided with knowledge about concretely usable technologies.[5] Mathematical formulas are therefore not subject to patent law.[6] In this context, the publication of the invention through a patent application or a patent plays a major role. Only through publication can the general public benefit from the described technical teaching. The consideration is a limited-term monopoly for the applicant and the naming of the inventor for the inventor.[7] A justification for patent law can also be seen in the fact that patent protection shields small and medium-sized enterprises against the dominance of large corporations.[8]

---

[1] Ann, § 3. Legal and economic policy assessment of patent protection in Ann, Patent Law 8th edition 2022, para. 37.

[2] Rogge/Melullis, Introduction Benkard, Patent Act: PatG, 12th edition 2023, para. 3–3b.

[3] *The Economist* v. 8.8.2015, A question of utility, p. 9; Time to fix patents, p. 43 ff.; Haedicke, § 1. Introduction, Haedicke/Timmann, Handbook of Patent Law, 2nd edition 2020, para. 12.

[4] *Reichman* 46 Houston Law Review 2009, 1115 (1117); Haedicke, § 1. Introduction, Haedicke/Timmann, Handbook of Patent Law, 2nd edition 2020, para. 15.

[5] Rogge/Melullis, Introduction, Benkard, Patent Act: PatG, 12th edition 2023, para. 1.

[6] § 2 Patent Act.

[7] Rogge/Melullis, Introduction in Benkard, Patent Act, 11th edition 2015, para. 1–2.

[8] Rogge/Melullis, Introduction, Benkard, Patent Act: PatG, 12th edition 2023, para. 1.

2.2 Functions of Patent Law

Several desirable functions or tasks are attributed to patent law. On the one hand, the reward of the inventor for their contribution to the further development of technology. Closely related to this is the incentive function, to further motivate the inventor to enrich technology. In addition, a patent realizes a property function. Here, a patent is regarded as proof of ownership of an intellectual creation. Intellectual creations are equated with tangible property. As with tangible property, the inventor must have the possibility to exclude unauthorized third parties from the use of his invention. Another function of patent law is described by the disclosure or contract theory, which assumes a "contract" between the inventor, who must publish his invention in a reproducible manner, and the state in order to obtain a state privilege in return.[9]

Patents often claim only minor improvements. The question may rightly be asked whether such minimal improvements deserve to be secured by an exclusive right with a state monopoly.[10] On the other hand, such small patented inventions also claim only a small scope of protection, so their impact on the market is likely to be minor.

Rewarding and incentivizing the inventor

The Patent Act is intended to reward the inventor and to encourage further inventions. The reward theory assumes that the inventor is entitled to a reward for his effort in creating the invention, which is to be provided by the state, since the state also benefits from the promotion of technological development. According to the incentive theory, something should be granted to the inventor so that he considers it an incentive to engage in inventive activity. The contract theory also assumes that something must be given to the inventor so that he discloses his invention to the public. The most cost-effective solution for the state is a time-limited monopoly.[11]

However, it should be noted that in most cases, the granting of a patent does not serve as an incentive or reward for the inventor, but rather for a company, since 90% of inventions are employee inventions, which are regularly claimed by the employer.[12] However, an incentive for the employee-inventor arises from the Employee Inventions Act, which provides for remuneration of the inventive employee after the claim to his invention by the employer.[13] The invention does not necessarily have to be filed for a patent for this purpose. An obligation to pay remuneration also arises if the invention is used as a trade

[9] BeckOK PatR/Fitzner, 29th ed. 15.7.2023, PatG § 1 Before §§ 1–25 (The Patent) para. 4–9.

[10] BeckOK PatR/Fitzner, 29th ed. 15.7.2023, PatG § 1 Before §§ 1–25 (The Patent) para. 10.

[11] Machlup, Die wirtschaftlichen Grundlagen des Patentrechts – 1. Teil, GRUR Ausl. 1961, 373, 377–379.

[12] Ann, § 3. Legal and economic policy assessment of patent protection in Ann, Patentrecht 8th edition 2022, para. 37.

[13] § 9 paragraph 1 Employee Inventions Act.

secret.[14] Patent law can therefore be regarded more as an entrepreneurial safeguard for R&D investments than as a direct incentive or reward for an inventor.

Investment Protection

With the protective function of the patent, the patent holder is granted a time-limited exclusive right of exploitation.[15] This is intended to enable the inventor to recoup the financial outlay for the investments made.[16] The protective function can therefore motivate companies to invest financial resources in research and development.

Through the patent with its protective function, which prevents imitation without authorization, the patented invention becomes a tradable economic asset that can be presented without concern. This significantly facilitates negotiations regarding licensing, as confidentiality agreements may be waived.[17]

Technical Informationto the public

Another objective of the Patent Act is to inform the public about new technologies.[18] In particular, this is intended to prevent uneconomical duplicate inventions.[19] The publication of patented inventions enables patentable further developments based on patent specifications, thereby promoting technological progress. The information function requires the protective function, so that only through clear legal protection of the inventor can he be motivated to inform the public.

Informing the public about the current state of technological development is a fundamental prerequisite for technological progress, since it is important for technological progress that what has already been achieved is made known.[20]

The necessity of the information function of the Patent Act is disputed. It is argued that technical inventions that are in use cannot be kept secret permanently anyway.[21] It can be countered that informing the public through patent publications at least does not cause any disadvantage, but rather provides an additional (accelerated?) publication in an information channel accessible to the public.

---

[14] § 17 Employee Inventions Act.

[15] G. Prosi, Wirtschaft und Wettbewerb 1980, p. 641.

[16] Nägerl/Neuburger/Steinbach, Artificial Intelligence: Paradigm Shift in the Patent System, GRUR 2019, 336, 341.

[17] Ann, § 3. Legal and economic policy assessment of patent protection in Ann, Patentrecht, 8th edition 2022, para. 44.

[18] Zitscher: On the Expansion of the Information Function of the Patent System GRUR 1997, 261.

[19] Häußer in The Management of Innovations, Staudt, E. (ed.), Frankfurt, 1986, p. 577; and IWD of 13. 5. 1993, p. 6.

[20] Deck, § 17 Protection against Imitation under Competition Law (§ 4 No. 3 UWG) in Hasselblatt, MAH, Industrial Property Protection, 6th edition 2022, para. 3.

[21] Ann, § 3. Legal and economic policy assessment of patent protection in Ann, Patentrecht 8th edition 2022, para. 31.

Informing the public is, however, significantly restricted by the 18-month confidentiality of patent applications. The introduction of the 18-month publication period was originally justified by the effort required for publication, which can no longer be maintained today given the possibilities of the Internet.[22]

Technological Progress

The fundamental task of patent law is the promotion of technological progress.[23] Technological progress occurs if a technical teaching of a patent leads to further development of technology.[24] The generation of technological progress can be interpreted as an indication of inventive step.

However, technological progress does not automatically mean that the respective invention is patentable. There are indeed inventions that represent significant technological progress but are nevertheless not based on the inventive step required for granting a patent.[25]

Technological progress is not a requirement for patentability.[26] An enrichment of technological progress is therefore not required for granting a patent. However, technically nonsensical inventions cannot be patented. According to the Patent Act, it may be sufficient for inventive step if an alternative solution is found, even if no progress is achieved.[27]

Although achieving technological progress is not a requirement for patentability, the possibility of patenting is nevertheless intended to encourage the achievement of technological progress.[28] Technological progress can be regarded as evidence of inventive

---

[22] Meitinger, The Disclosure of the Patent Application after 18 Months: Is This Still Up-to-Date? Mitteilungen der deutschen Patentanwälte, 2017, July/August, 303.

[23] Plant, Economica 1934, and Silberston, Lloyd's Bank Review 1967.

[24] Einsele, PatG § 4[Invention based on inventive step] in BeckOK Patentrecht, Fitzner/Lutz/Bodewig, 26th Edition, as of 15.10.2022, para. 56.; § 4 sentence 1 Patent Act or Article 56 EPC. BGH BlPMZ 1966, 164 – Soup recipe; GRUR 1972, 704 – Water treatment; GRUR 1972, 707 – Adjustable stretching roller; Mitt. 1978, 136 – Oil pipe; BlPMZ 1966, 208 – Appetite suppressant I; BlPMZ 1969, 251 – Disiloxane; BlPMZ 1971, 131 – Anthradipyrazole; BlPMZ 1972, 319 – Imidazolines; GRUR 1964, 676 – Lapping; Mitt. 1972, 235 – Frosted candle; GRUR 1967, 590 – Garage door; BlPMZ 1976, 192 – Alkylene diamines.

[25] No inventive step despite significant technological progress: BGH GRUR 1969, 182 – Concrete dispenser.

[26] In § 1 of the Patent Act of 1968, achieving technological progress was still required in order to obtain a patent.

[27] BGH GRUR 2015, 983 para. 31 – Aircraft condition; Mes, PatG § 1 [Requirements for grant] in Mes, Patentgesetz Gebrauchsmustergesetz, 5th edition 2020, para. 93.

[28] BGH GRUR 1996, 109 – Clinical trials I; Einsele, GebrMG § 1 [Requirements for protection], BeckOK Patentrecht, Fitzner/Lutz/Bodewig, 26th Edition, as of 15.10.2022, para. 51.

step.[29] However, an invention can be new and inventive, and thus patentable, even if the invention is not advantageous compared to the prior art.[30]

Consumer Welfare

One task of patent law is to ensure socially beneficial inventions.[31] Social utility can be an indication of inventive step.[32] Patent law therefore also aims to increase consumer welfare.[33]

It has been controversial from the outset whether patent law can serve consumer welfare at all. The first Nobel laureate in physics, Wilhelm Conrad Röntgen, deliberately refrained from patenting applications of his "special radiation," now known as X-rays, in order to make the benefits of his discovery accessible to people without legal obstacles. Röntgen was fully aware of the technical application possibilities and the fundamental patentability.[34]

2.3 Invention

In patent law, there is the central concept of the "invention." Only for an object to which the concept of "invention" applies is the Patent Act accessible. The Patent Act does not contain a legal definition of the term invention. The Patent Act only requires that an invention has a technical character.[35]

The legislator has not defined when a technical character or a patentable invention is present. This "gap" was intentionally left by the legislator. The absence of a definition of the invention is intended to allow this concept to be adapted to the respective state of the art.[36] The legislator has therefore assigned it to legal practice, in particular to the Patent

---

29 BGH GRUR 1996, 757 – Gear cutter; BPatG GRUR 1995, 397– Exterior mirror arrangement.

30 BGH GRUR 2015, 983 – Aircraft condition.

31 BGH GRUR 1956, 77 – Binding wire; BGH GRUR 1955, 29 – Nobelt bundle; BGH GRUR 1996, 109 – Clinical trials I; Einsele, GebrMG § 1 [Requirements for protection] in BeckOK Patentrecht, Fitzner/Lutz/Bodewig, 26th Edition, as of 15.10.2022, para. 50.

32 BPatG GRUR 1995, 397 – Exterior mirror arrangement; Einsele, GebrMG § 1 [Requirements for protection] in BeckOK Patentrecht, Fitzner/Lutz/Bodewig, 26th Edition, as of 15.10.2022, para. 50.

33 Joseph Straus, Patent application as abuse of dominant market position under Article 82 EC Treaty? GRUR Int 2009, 93, 99.

34 Beyer: "Patent and ethics in the mirror of technical evolution" GRUR 1994, 541.

35 Section 1(1) Patent Act; Beyer, Patent and Ethics in the Mirror of Technical Evolution, GRUR 1994, 541, para. 543.

36 See Bundesratsdrucksache No. 14 1876/77 regarding Section 1; Becker refers to "grey areas for the evolution of legal rules" (Becker, Maximilian: On the Freedom to Act Illegally, ZUM 2019, 636, 642). The legal concept of the invention represents such a nucleus for the evolution of the Patent Act.

Office and the courts, as a task, to define for which products and processes the Patent Act should be accessible.

A current example of the development of the meaning of the term invention is software patents.[37] In this area, a development of the law by the patent offices and courts towards a more liberal view can be observed, so that software is increasingly patentable.[38]

Pure discoveries, scientific theories, mathematical methods, schemes and rules for games and business activities are excluded from patent protection.[39] Important for a startup: a business model cannot be protected by a patent. In addition, methods for surgical or therapeutic treatment of humans or animals are also explicitly excluded from patent protection.[40]

2.4 Patent Application

An invention can be applied for as a patent.[41] A patent application does not automatically lead to an exclusive right. As long as the application has not resulted in a grant, the invention can be used by anyone. However, the applicant is entitled to compensation from everybody who uses the submitted invention.[42] The compensation claim is not damages as in the case of a patent infringement and is also to be set significantly lower.[43]

2.5 Patent

A patent application whose subject matter is new, inventive, and industrially applicable can be granted as a patent.[44] With a patent, any unauthorized person can be prohibited from any kind of use of the patent-protected technical teaching. Patent-protected uses are in particular the manufacturing, offering, and marketing of the protected invention.[45]

---

[37] Lutz van Raden, the Informatics Dove – Considerations on the patentability of information-related inventions, GRUR 1995, 451.

[38] Andreas Wiebe describes the history of software patenting and refers to a "long ordeal" (Andreas Wiebe, Patent Protection and Software Development – an Irreconcilable Contradiction? From the book "Open Source Yearbook 2004 – Between Software Development and Social Model", 2004, pp. 277–292).

[39] Section 1(3) No. 3 Patent Act.

[40] Section 2a(1) No. 2 Patent Act.

[41] Section 34(1) Patent Act.

[42] Section 33(1) Patent Act.

[43] Section 139(2) sentence 1 Patent Act.

[44] Section 1(1) Patent Act.

[45] Section 9 No. 1 Patent Act.

In addition to patent infringement, where all features of a claim are fulfilled (direct patent infringement), an indirect patent infringement can be asserted. The prerequisite is that the patent infringement relates to essential elements of the patent-protected invention which are also suitable for use of the invention. In addition, the suitability must have been known or at least obvious to the infringer. However, in the case of indirect patent infringement, only offering and supplying within the country can be asserted. The type of use of manufacturing cannot lead to indirect patent infringement.[46]

2.6 Inventor

Theinventor is the owner of his invention.[47] The applicant is the natural or legal person who applies for the invention as a patent. If the applicant is not the inventor or the invention has not been assigned to him, he acts without authorization and there is an unlawful taking (vindication). An assignment of the invention can be made by contract or by the claiming of the invention of an employee under the Employee Inventions Act by his employer.

The inventor must be a natural person. Previously, it was possible to file so-called "company inventions" without naming an inventor. Some abuse occurred with company inventions, and inventors were deprived of their inventorship.[48]

2.7 Applicant

TheApplicant is a natural or legal person who files the invention for a patent and is the holder of the intellectual property right. The applicant is only entitled to file the application if he is the inventor or has acquired the invention from the inventor by law or contract. The patent office is not able to verify the entitlement of the applicant. In order not to delay the proceedings before the patent office, the patent office initially assumes, unless there are indications to the contrary, that the applicant is entitled.[49] The applicant must designate the inventor within fifteen months after the filing date and indicate how the right to the invention was transferred to him, if he is not the inventor.[50] The information is not verified by the patent office.[51]

[46] Section 10(1) Patent Act.

[47] Section 6 sentence 1 Patent Act.

[48] Mediger, GRUR 1952, 67.

[49] Section 7(1) Patent Act.

[50] Section 37(1) sentence 1 Patent Act.

[51] Section 37(1) sentence 3 Patent Act.

In particular, a claim to the invention of an employee by his employer can result in a transfer of rights.[52] A transfer of rights by law occurs if the employer does not actively release the invention within four months after invention disclosure.[53] In this case, a claim is legally presumed.

If the applicant is not the inventor and no transfer of rights has taken place, the applicant is acting without authorization. In this case, an unlawful appropriation (vindication) exists, and the entitled party may demand from the applicant that the application or the patent be transferred to him.

The party from whom the invention was unlawfully appropriated may enforce his rights by filing an opposition or an action for nullity. In particular, the entitled party may, by means of an opposition[54] or an action for nullity[55] have the unlawfully acquired patent declared invalid. Alternatively, the entitled party has the option to file a vindication action.[56]

2.8 Patentability Requirements

The independent claims of a patent application must be novel, based on an inventive step, and be industrially applicable in order for the claimed invention to be patentable.[57] The independent claims are those that do not refer back to preceding claims.

Novelty

A claim is novel if it does not form part of the state of the art.[58] The state of the art comprises all knowledge that was made available to the public before the filing or priority date. Disclosure may have occurred in any form, for example in written or oral form, by use or in any other way.[59]

Patent applications that were filed with the patent office before the filing or priority date, but were only published on or after the filing or priority date can also be prejudicial to novelty.[60] This so-called "subsequently published" state of the art, however, is not taken into account when assessing inventive step, as it could not have been combined with other documents.[61]

---

[52] Section 7(1) Employee Inventions Act.

[53] Section 6(2) Employee Inventions Act.

[54] Section 59(1) sentence 1 in conjunction with Section 21(1) no. 3 Patent Act.

[55] Section 81(1) sentence 1 in conjunction with Sections 22, 21(1) no. 3 Patent Act.

[56] Section 8 Patent Act.

[57] Section 1(1) Patent Act.

[58] Section 3(1) sentence 1 Patent Act.

[59] Section 3(1) sentence 2 Patent Act.

[60] Section 3(2) Patent Act.

[61] Section 4 sentence 2 Patent Act.

Inventive Step

A person skilled in the art will not apply a technical teaching rigidly, but will adapt it based on their technical knowledge and skill. Modifications of technical teachings are therefore common practice. If such modifications were already patentable, patent law would be more harmful than beneficial, as it would hinder the normal activities of a skilled person.[62] For this reason, all modifications of the existing state of the art that are obvious to the skilled person based on their technical knowledge and skill are not eligible for protection.[63]

The assessment of inventive step is based on a hypothetical average skilled person who would be entrusted with development activities in the relevant technical field.[64] The skilled person possesses technical knowledge and skill. Technical knowledge includes both field-specific knowledge and general technical knowledge.[65] The skilled person can thus also "read into" a patent specification aspects that are not explicitly mentioned.

The technical skill of the patent law skilled person has been acquired through work routine and personal experimentation. The hypothetical average skilled person in patent law is therefore to be credited with a creative achievement, which, however, does not reach the level of inventive step as required by the Patent Act.[66]

So-called subsequently published state of the art refers to applications that were filed with the patent office before the filing or priority date and were only published by the patent office on or after the filing or priority date. This special state of the art is not considered in the assessment of inventive step, as it could not have been combined with other documents of the state of the art and therefore may call novelty into question, but not inventive step.[67]

Could-Would Test

A claim does not lack inventive step merely because it results from any arbitrary combination of documents from the state of the art. In addition, the skilled person must also have had a motivation to combine these documents.[68] If such motivation is absent, the combination of the documents was not obvious and it was inventive to combine exactly these documents.

---

[62] Ann, Section 18. Inventive Achievement in Ann, Patent Law, 8th edition 2022, para. 2.

[63] Section 4 sentence 1 Patent Act.

[64] Benkard PatG/Asendorf/Schmidt/Tochtermann, 12th ed. 2023, PatG § 4 para. 72.

[65] Benkard PatG/Asendorf/Schmidt/Tochtermann, 12th ed. 2023, PatG § 4 para. 69.

[66] Section 4 sentence 1 Patent Act; BGH GRUR 1986, 372 – Platelet Counting.

[67] Section 4 sentence 2 Patent Act.

[68] BGH GRUR 2010, 407 single-piece eyelet.

The Could-Would Test is intended in particular to exclude ex post considerations that do not do justice to the situation of the inventor at the time of creating the invention, who did not yet know the way to the creation of the invention.[69]

Industrial Applicability
This criterion is almost always met, as any possibility of use of the subject matter of the relevant claim in any industrial field is sufficient.[70] Only a product that is completely useless in every respect could fail to meet the criterion of industrial applicability.

2.9 Grant Procedure

The German patent office conducts a formal examination immediately after receipt of the application documents to determine whether a filing date can be accorded to the documents. A substantive examination, at the positive conclusion of which the grant of the patent stands, is only initiated by the patent office after a request for examination has been filed. Instead of a request for examination, a request for search may be filed, which results in a search for the relevant state of the art, without the patent office making an assessment of the subject matter of the application in light of the searched state of the art.

Formal Examination
The patent office conducts a formal examination of the application documents to determine whether the application contains a request for grant of a patent, an indication of the applicant, and the other components of a patent application.[71] In addition, the application documents must contain an abstract.[72] If necessary, the patent office will request the applicant within a specified period to remedy any deficiencies. If the deficiencies are not remedied within the period, the application will be rejected.[73]

However, if the subject matter of the application is excluded from patentability, the applicant will be informed and given the opportunity to respond within a specified period.[74] If the applicant cannot convince the Patent Office otherwise, the application will be rejected.[75]

---

[69] EPO T 47/91; Singer/Stauder/Luginbühl Art 56 EPC para. 56; on their inadmissibility BGH GRUR 2001, 232 hole punch.

[70] BGH 1985, 117 – Obviousness Test.

[71] Section 42(1) sentence 1 in conjunction with Section 34 Patent Act.

[72] Section 42(1) sentence 1 in conjunction with Section 36 Patent Act.

[73] Section 42(3) sentence 1 Patent Act.

[74] Section 1 paragraph 3 or Section 2 Patent Act; Section 42 paragraph 2 Patent Act.

[75] Section 42 paragraph 3 sentence 1 Patent Act.

Request for Search

The applicant may file a request for a search.[76] The Patent Office will then determine the prior art relevant for examining patentability. If it turns out that the set of claims describes several independent inventions, the search will be based on the first mentioned invention.[77]

Request for Examination

The applicant, and any third party, may file a request for examination. In this case, not only will the prior art relevant for assessing grantability be determined, but the Patent Office will also communicate its assessment of grantability in the form of an office action. The request for examination must be filed within a period of seven years from the filing date. Otherwise, the application will be rejected.[78]

If the request for examination is filed simultaneously with or at least shortly after the submission of the application documents, the Patent Office will endeavor to provide the first office action to the applicant within the priority period of one year. The applicant can then decide, based on the opinion of the Patent Office regarding the patentability of the application, whether subsequent filings abroad are advisable.

Examination Procedure

In the examination procedure, it is examined whether a set of claims can be developed that is novel and inventive. Typically, two to three office actions from the Patent Office must be answered before the legal situation is clarified. At that point, allowable claim formulations are available or the application is rejected.

The applicant should, as a precaution, submit a request for oral proceedings together with his response to the office action from the Patent Office.[79] If no allowable claim formulations can be found in the written procedure, there is still a chance to achieve a grant of the patent in the oral proceedings.

The patent grant procedure currently takes between three and eight years. An applicant should therefore not wait for the grant of the patent before starting to exploit the invention. A much better strategy is to begin commercializing the invention immediately after filing the application documents. In this case, the applicant will have gained an idea of the economic significance of his invention after just a few months and can decide whether subsequent filings abroad claiming the priority of the first application are worthwhile. In doing so, the one-year priority period should not be forgotten.

---

[76] Section 43 paragraph 1 sentence 1 Patent Act.

[77] Section 43 paragraph 6 in conjunction with Section 34 paragraph 5 Patent Act.

[78] *Section 44 paragraph 2 sentence 1 Patent Act.*

[79] Section 46 paragraph 1 sentence 3 Patent Act.

Third-Party Submissions
The examination procedure takes place between the applicant and the Patent Office. No other parties are involved. However, any third party is permitted to provide the Patent Office with documents that may question the patentability.[80]

Applicant Appeal
An appeal against a decision of rejection by the Patent Office may be filed as a complaint before the Federal Patent Court.[81] The complaint may be filed within a period of one month upon receipt of the decision at the German Patent and Trade Mark Office to contest.[82] The examining division that issued the rejection is given the opportunity to remedy the situation. If no remedy is provided, the complaint is forwarded to the Federal Patent Court without comment from the Patent Office.[83]

2.10 Opposition

With an opposition, the legal validity of a granted German patent can be reviewed. Any person may file an opposition. There is no need to demonstrate a specific interest in legal protection.

Opposition period
An opposition may be filed within a nine-month period after publication of the grant of the patent at the Patent Office. The opposition must be substantiated with reasons.[84]

Grounds for opposition
A patent is revoked in particular if it is not novel and inventive, or if its subject matter is fundamentally not patentable.[85] Another ground for revocation exists if the protected technical teaching is not executable by the skilled person.[86] Furthermore, an inadmissible extension exists if the subject matter of the claims cannot be derived from the originally filed application documents.[87]

[80] *Section 43 paragraph 3 sentence 2 Patent Act.*
[81] Section 47 paragraph 2 sentence 1 Patent Act.
[82] Section 73 paragraph 2 sentence 1 Patent Act.
[83] Section 73 paragraph 3 sentence 3 Patent Act.
[84] Section 59(1) sentence 2 Patent Act.
[85] Section 21(1) no. 1 in conjunction with Sections 1 to 5 Patent Act.
[86] Section 21(1) no. 2 Patent Act.
[87] Section 21(1) no. 4 Patent Act.

The ground for revocation due to unlawful removal can only be asserted by the injured party themselves.[88] Unlawful removal or vindication exists if an unauthorized person has filed the invention for patent protection.[89]

Procedure

The opposition proceedings take place before a patent division of the German Patent Office.[90] The opposition proceedings begin with a written phase, during which the parties exchange documents regarding the legal validity of the contested patent. If one of the parties has requested an oral hearing or if the Patent Office considers it appropriate, an oral hearing will be held before the patent division makes its decision.[91] An appeal against the decision of the Patent Office may be lodged with the Federal Patent Court.

Decision

The decision on the opposition is made by resolution of the patent division.[92] The contested patent may be completely revoked or maintained in a limited form. If the opposition is unsuccessful, the contested patent remains in force in its entirety. Limited maintenance may result from an amendment of the claims, the description, or the drawings.[93] The decision of the opposition division has retroactive effect. A revoked patent is therefore deemed never to have had any effect.[94]

Intervention in the opposition proceedings

A third party against whom patent infringement proceedings are pending may intervene in ongoing opposition proceedings, even if the opposition period has already expired. The intervention may be declared up to a period of three months after an action has been brought against the third party by the patent proprietor.[95]

Ex officio investigation principle

The patent division can continue the opposition proceedings if the opponent withdraws their opposition. This does not apply if the sole ground for revocation is unlawful removal. In this case, the proceedings will be definitely terminated.[96]

---

[88] Section 59(1) sentence 1 Patent Act.

[89] Section 21(1) no. 3 Patent Act.

[90] Section 61(1) sentence 1 Patent Act.

[91] Section 59(3) sentence 1 Patent Act.

[92] Section 61(1) sentence 1 Patent Act.

[93] Section 21(2) sentence 2 Patent Act.

[94] Section 21(3) sentence 1 Patent Act.

[95] Section 59(2) sentence 1 Patent Act.

[96] Section 61(1) sentences 3 and 4 Patent Act.

Costs
In opposition proceedings, each party bears its own costs. The opponent must also pay an opposition fee. The principle of defeat, under which the losing party must bear the costs of the opposing party, does not apply in opposition proceedings.[97]

2.11 Nullity

If the opposition period has already expired, action can only be taken against a granted German patent by means of a nullity action before the Federal Patent Court. There is no deadline to be observed for filing a nullity action. However, filing a nullity action before the expiry of the opposition period is excluded.[98]

Grounds for Nullity
The grounds for revocation correspond to those of the opposition procedure. A patent is declared null and void if it is not novel and inventive, or if its subject matter is fundamentally not patentable.[99] Another ground for revocation exists if the protected technical teaching is not executable by the person skilled in the art.[100] An inadmissible extension exists if the subject matter of the claims cannot be derived from the originally filed application documents.[101] Additionally, a patent can be declared null and void if it was inadmissibly extended during an opposition procedure.[102] If essential content of the contested patent was unlawfully taken, this also constitutes a ground for nullity.[103]

Contents of a nullity action
The nullity action is brought by filing the statement of claim. The statement of claim must indicate the plaintiff, the defendant, the contested patent, and specify to what extent the patent is being contested or for which claims of the patent a declaration of nullity is being requested. The action must be substantiated with reasons, and the facts and documents on which the grounds for revocation are based must be attached. If essential parts of the action are missing, the Federal Patent Court will request the plaintiff to submit these parts. A deadline will be set for this.[104]

---

[97] Section 62 Patent Act.

[98] Section 81 paragraph 2 sentence 1 Patent Act.

[99] Section 22 paragraph 1 in conjunction with Section 21 paragraph 1 no. 1 in conjunction with Sections 1 to 5 Patent Act.

[100] Section 22 paragraph 1 in conjunction with Section 21 paragraph 1 no. 2 Patent Act.

[101] Section 21 paragraph 1 no. 4 Patent Act.

[102] Section 22 paragraph 1 in conjunction with Section 21 paragraph 1 no. 4 Patent Act.

[103] Section 22 paragraph 1 in conjunction with Section 21 paragraph 1 no. 3 Patent Act.

[104] Section 81 paragraph 5 sentences 1 to 3 Patent Act.

Procedure

The Federal Patent Court serves the statement of claim on the defendant and gives him the opportunity to object within a period of one month.[105]

If no objection is raised by the defendant, the facts asserted by the plaintiff are deemed proven and the Federal Patent Court decides immediately without an oral hearing.[106]

If the defendant objects, he may, within a period of two months after receipt of the action, substantiate their objection. The objection is communicated to the plaintiff.[107]

The presiding judge of the nullity panel sets a date for the oral hearing, which is communicated to the parties. An oral hearing may be dispensed with if the parties agree.[108]

The nullity panel may, within a period of six months after receipt of the action, give a judicial indication to the parties as to which points still need to be clarified and, in particular, which should be discussed at the oral hearing. To prepare the judicial indication, the nullity panel may give the parties an opportunity to comment within a set period.[109]

The parties may respond to the judicial comment with their own written submission to clarify their, in particular differing, legal opinion.[110]

If means of attack and defense, an amendment of the action, or an amended version of a set of claims are submitted only after a period set by the nullity panel for comment, they may be rejected as too late.[111]

Decision

The nullity panel of the Federal Patent Court decides by judgment.[112] The decision has "ex tunc" effect, meaning from the outset. A patent declared null and void was therefore never legally effective at any time.[113]

Costs

Before the Federal Patent Court, as before an ordinary court, the loser pays principle applies. The losing party must therefore bear the costs of the prevailing party and the court costs.[114]

---

[105] Section 82 paragraph 1 Patent Act.

[106] Section 82 paragraph 2 Patent Act.

[107] Section 82 paragraph 3 sentences 1 and 2 Patent Act.

[108] Section 82 paragraph 4 sentences 1 and 2 Patent Act.

[109] Section 83 paragraph 1 sentences 1, 2 and 4 Patent Act.

[110] Section 83 paragraph 2 sentence 1 Patent Act.

[111] Section 83 paragraph 4 Patent Act.

[112] Section 84 paragraph 1 sentence 1 Patent Act.

[113] Section 22 paragraph 2 in conjunction with Section 21 paragraph 3 sentence 1 Patent Act.

[114] Section 84 paragraph 2 sentence 2 Patent Act.

2.12 Complaint, Appeal, and Appeal to a Higher Court

The complaint is the legal remedy used to contest a decision of the German Patent Office. By means of a legal complaint before the Federal Court of Justice, a complaint proceeding can be reviewed. The nullity appeal before the Federal Court of Justice is the appellate instance for a nullity proceeding before the Federal Patent Court.

Complaint before the Federal Patent Court

The complaint before the Federal Patent Court serves to challenge a decision of the Patent Office, for example, the rejection of a patent application or an opposition.[115] The complaint must be filed within one month after receipt of the decision at the Patent Office.[116]

The complaint is first forwarded to the office or department of the Patent Office that issued the contested decision. If the relevant office of the Patent Office determines that the complaint is justified, it will amend its decision. In this way, the complaint can be "remedied".[117] If the complaint is not remedied, the complaint is forwarded to the Federal Patent Court without a statement of grounds from the Patent Office.[118]

The President of the Patent Office is permitted, if in his view this is necessary to safeguard the public interest, to submit written statements to an complaint proceeding before the Federal Patent Court.[119] The President of the Patent Office may alternatively, if legal questions of fundamental importance are raised in the complaint proceeding, join the complaint proceeding as a party, provided the Federal Patent Court permits this.[120]

The decision concludes the complaint proceeding.[121] A complaint that is inadmissible will be dismissed without an oral hearing.[122]

Alternatively, the Federal Patent Court may set aside the contested decision and refer the matter back to the Patent Office without deciding on the merits. This occurs if the contested procedure before the Patent Office has a material defect or if new facts or evidence are present that are considered highly relevant to the contested decision.[123] The Patent Office must take into account the legal assessment of the Federal Patent Court, which led to the setting aside, in its new decision.[124]

---

[115] Section 73(1) Patent Act.
[116] Section 73(2) sentence 1 Patent Act.
[117] Section 73(3) sentence 1 Patent Act.
[118] Section 73(3) sentence 3 Patent Act.
[119] Section 76 sentence 1 Patent Act.
[120] Section 77 sentence 1 Patent Court.
[121] Section 79(1) Patent Act.
[122] Section 79(2) Patent Act.
[123] Section 79(3) sentence 1 nos. 2 and 3 Patent Act.
[124] Section 79(3) sentence 2 Patent Court.

Legal Complaint

A legal complaint serves the legal review of an complaint proceeding before the Federal Court of Justice. However, the Federal Patent Court must have admitted the legal complaint, or the complaint proceeding must have material defects, for example, violation of the right to be heard or the absence of reasons in the decision of the Federal Patent Court.[125]

The Federal Patent Court will admit the legal complaint if the complaint proceeding involved a matter of fundamental importance. This is intended to ensure uniform case law.[126]

Appeal

The appeal before the Federal Court of Justice is the legal remedy for the review of a nullity judgment of the Federal Patent Court.[127] An appeal must be filed with the Federal Court of Justice within one month after receipt of the judgment of the Federal Patent Court.[128]

The nullity appeal is an appellate instance. New facts are not admitted. A nullity appeal serves solely to assess whether the Nullity Senate of the Federal Patent Court correctly applied the legal provisions.[129]

2.13 Effect and Limits of Patent Protection

A German patent grants his patent holder a claim for injunctive relief against any patent infringement. For this, a threat of first-time infringement is sufficient, for example, if it has already been announced that the patent will not be respected, or a risk of recurrence, since the patent has already been infringed in the past. In the case of fault, which is generally assumed, the patent holder can claim damages from the infringer.

Claim for Injunctive Relief

A claim for injunctive relief arises if all features of a claim are realized by a product or process of an unauthorized party (direct patent infringement)[130] or already if essential features are fulfilled (indirect patent infringement)[131]. Fault, i.e., intent or negligence, is not a prerequisite for the claim for injunctive relief.

[125] Section 100(1) Patent Act.
[126] Section 100(2) Patent Act.
[127] Section 110(1) Patent Act.
[128] Section 110(3) sentence 1 Patent Act.
[129] Section 111(2) in conjunction with Section 111(1) Patent Act.
[130] Section 139(1) sentence 1 in conjunction with Section 9 Patent Act.
[131] Section 139(1) sentence 1 in conjunction with Section 10 Patent Act.

Other Claims

As a rule, in the case of a patent infringement fault, i.e., intent or negligence, is assumed, so that the patent holder is regularly entitled to a claim for damages.[132] In addition, the patent holder can assert a claim for recall and demand that infringing products be permanently removed from the distribution chain.[133] Additionally, a claim for information exists, so that the patent holder must be informed about the distribution channels and the origin of the infringing products.[134] The patent holder is entitled to a claim for destruction. The patent holder can therefore demand the destruction of infringing products. However, the legitimate interests of third parties must be taken into account.[135] Furthermore, the patent holder can demand the surrender of anything that the infringer has obtained through use of the patent at the expense of the patent holder.[136]

Entitlement to Assert Claims

The claims arising from a patent can be asserted by the patent holder and an exclusive licensee. A simple licensee does not have standing to sue, but can be authorized through voluntary procedural standing.[137]

Limits of Patent Protection

Private use cannot be prevented by the patent holder.[138] The individual preparation of a medication by a pharmacist is also outside the scope of protection of a patent.[139]

The patent holder is required to describe the technical teaching of their patent in the application in such a way that it can be carried out by a person skilled in the art. Furthermore, the patent holder is obliged to allow the patent office to make the patented technical teaching accessible to the public. It is therefore only logical that an experimental use exemption exists, according to which anyone is permitted to subject the patent holder's technical teaching to experiments in order to find out more about the invention.[140] However, if the experiments are not aimed at researching the patented invention, they can be prohibited by the patent holder.[141]

---

[132] Section 139(2) sentence 1 Patent Act.

[133] Section 140a(3) sentence 1 Patent Act.

[134] Section 140b(1) Patent Act.

[135] Section 140a Patent Act.

[136] Section 812(1) sentence 1 BGB..

[137] Mes, 5th ed. 2020, PatG § 139 para. 44–51; Osterrieth, Part 6. Patent Infringement Osterrieth, Patent Law, 6th edition 2021, para. 927.

[138] Section 11 no. 1 Patent Act.

[139] Section 11 no. 1 Patent Act.

[140] Section 11 no. 2 Patent Act.

[141] Federal Court of Justice 11.7.1995 – X ZR 99/92, GRUR 1996, 109 – Clinical Trials I; Federal Court of Justice 17.4.1997 – X ZR 68/94, NJW 1997, 3092 – Clinical Trials II.

In addition, a right of prior use must be observed for anyone who had already used the technical teaching of the invention or had made serious preparations to use it at the time of filing or priority.[142]

2.14 Infringement Proceedings

A warning notice is the last opportunity to resolve a patent infringement out of court. If the legal situation is unclear, it is advisable to first address a request for justification to the potential infringer. If urgency exists, for example due to an upcoming trade fair, a preliminary injunction can be sought. Otherwise, legal action before an ordinary court remains.

Request for Justification

If the patent holder is not certain that an actual patent infringement has occurred, a request for justification can initially be addressed to the potential infringer for legal clarification. This gives the potential infringer the opportunity, for example, to point out a prior use right. In addition, attention can be drawn to features of the allegedly infringing products that would exclude a patent infringement.

Warning Letter

A warning letter must make it clear to an infringer that this is the last opportunity to resolve a patent infringement out of court and that if the attempt at warning fails, legal action will be taken.

A warning letter must include a request to submit a cease-and-desist declaration subject to penalty. Typically, a pre-formulated cease-and-desist declaration is attached to the warning letter, which the infringer only needs to sign and return.

The cease-and-desist declaration must include the agreement of a contractual penalty, which must be sufficiently high to demonstrate the seriousness of the signatory's intention to refrain from the patent infringement in the future.

A warning letter is not a prerequisite for either a preliminary injunction or legal proceedings. Legal action for patent infringement can be brought immediately before a civil court. However, if the defendant immediately acknowledges the claim, the plaintiff must bear the costs of the proceedings.

Preliminary Injunction

If it is important that a judicial decision is made quickly, because otherwise irreparable damage would occur, a preliminary injunction can be requested.[143] Typical examples

[142] Section 12(1) sentence 1 Patent Act.

[143] §§ 935, 940 ZPO.

include trade fairs that need to be cleared of infringing products. A decision after the end of the trade fair, stating that infringing products should not have been exhibited, will be of little significance to the patent holder.

Urgency is therefore an essential prerequisite for the issuance of a preliminary injunction. In addition, the judge must be provided with sufficient certainty regarding the legal validity of the underlying patent. With a patent application or a utility model that has not even successfully passed an official examination procedure, it will be difficult to obtain a preliminary injunction. If, in addition, the patent infringement is not immediately apparent to the judge from the submitted documents, he will hardly issue a preliminary injunction without hearing the opposing party.

Therefore, if the legal validity of the patent has not been confirmed by a completed opposition or nullity proceedings and the patent infringement is rather difficult to assess, at least a short-term oral hearing will be scheduled before a decision is made on the preliminary injunction.

Given the fact that patent infringements often revolve around complex technical matters, issuing a preliminary injunction without hearing the opposing party is the exception.

Legal Proceedings

Patent litigation cases are heard before specialized patent litigation chambers of selected regional courts at first instance and higher regional courts at second instance.[144] For each federal state, regional and higher regional courts have been selected to enable specialization in patent litigation cases. These courts have exclusive jurisdiction for patent infringement proceedings in their respective federal state.[145]

The patent infringement procedure comprises several steps. First, the patent holder drafts a statement of claim, in which the allegedly infringing product is compared with the features of the patent claims. The statement of claim must in particular make clear that the allegedly infringing product falls within the scope of protection of a claim.

The statement of claim is filed with the regional court. The regional court serves the statement of claim on the alleged infringer and sets a deadline for a response. The defendant submits his defense to the regional court, and the plaintiff has the opportunity to reply to the defense with a rejoinder. The defendant may then respond to the rejoinder with a surrejoinder.

By exchanging written pleadings, the disputed matter is examined in detail, so that the court is provided with a comprehensive explanation from both parties. After the written stage, the parties and the court are very well prepared for the subsequent oral hearing. After the close of the oral hearing, the court pronounces its judgment.

[144] § 143 paragraph 1 Patent Act.

[145] § 143 paragraph 2 Patent Act.

The decision of the regional court can be appealed by an appeal to the higher regional court. A revision procedure against the decision of the higher regional court before the Federal Court of Justice is only permitted in rare exceptional cases.

Representation by a lawyer is mandatory before the regional and higher regional courts, so a lawyer is required for representation. A patent attorney does not fulfill the requirement for legal representation before the ordinary courts (regional and higher regional courts and the Federal Court of Justice).[146]

2.15 Interpretation of Claims

The startup founder should understand the value of a patent. The value of a patent results from its scope of protection, i.e., from the subject matter whose use can be prohibited to a third party on the basis of the patent.

The startup founder must therefore be able to understand the claims of a patent. From this understanding, he can, on the one hand, determine whether his own products infringe third-party patents. In addition, this knowledge enables him to assess the value of his own patents.

In order to properly understand the scope of protection of a patent, the independent claims must be "interpreted." A correct understanding corresponds to the understanding of a competent court that must make a decision based on a patent claim. "Interpretation" means that the words of the claim are to be construed as they would be understood by a person skilled in the art.

The interpretation of a legal provision is a typical legal activity. Before a legal provision is applied, it must necessarily be interpreted. Lawyers have developed several variants for this purpose, which create an understanding of the relevant legal provision from different perspectives. Normally, all variants of interpretation are applied in order to arrive at a convincing interpretation of the legal provision from an overall perspective.

The patent claims of a patent can be regarded as legal provisions, since by law they determine which uses are inadmissible. Accordingly, patent claims must also be interpreted in order to correctly determine the statutory scope of protection.

An interpretation of the claims is required even if the wording appears to be clear. Reference to the fact that a patent specification is its own lexicon illustrates the indispensable necessity of interpretation in the context of the entire disclosure of the patent specification.

The interpretation is intended to ensure that the patent proprietor receives appropriate protection and that legal certainty is provided through a predictable and reproducible

[146] § 78 paragraph 1 sentence 1 ZPO.

interpretation of the claims.[147] In particular, the interpretation must serve to eliminate ambiguities and to determine the meaning of individual terms.[148] Finally, the interpretation must make it possible to determine the various embodimentsof the invention that are covered by the claim.

The scope of protection of a patent is determined by the wording of the patent claims.[149] However, the entire patent specification must be used for interpretation.[150] Even if a claim appears clear and unambiguous, the description and the drawings of the patent specification must be taken into account for interpretation.[151] Interpretation is particularly necessary if individual terms of the claim are unclear.[152]

However, it should not be mistakenly assumed that the description could create its own scope of protection. Nor do the claims merely indicate in general terms what is claimed by the patent. Rather, the claims determine the scope of protection.[153] The description and drawings serve only to properly understand the claims. Claims, description, and drawings do not constitute sources of equal rank for determining the scope of protection of a patent. If there are contradictions between the claims and the description and the drawings, the claims must always take precedence in determining the scope of protection.

The independent claims each define their own scope of protection. The independent claims are therefore to be regarded as independent with respect to the scope of protection. It may even occur that identical terms in the independent claims are to be understood differently.

The independent claims are the main claim and the co-ordinate claims. The main claim is the claim that is mentioned first. The co-ordinate claims are claims that do not refer to preceding claims. Usually, the independent claims belong to different categories of patents. Patent categories are, in particular, apparatus and method claims.

[147] Article 1 of the Protocol on the Interpretation of Article 69 EPC as amended by the Act revising the EPC of 29 November 2000, https://www.epo.org/law-practice/legal-texts/html/epc/2020/d/ma2a.html, accessed on 1.9.2022.

[148] Federal Court of Justice (BGH), Judgment of 12 March 2002 – X ZR 135/01, GRUR 2002, 519 – Schneidmesser II.

[149] Section 14 sentence 1 Patent Act or Article 69(1) sentence 1 EPC: "The scope of protection of the (European) patent and the (European) patent application shall be determined by the patent claims."

[150] Section 14 Patent Act.

[151] Section 14 sentence 2 Patent Act or Article 69(1) sentence 2 EPC: "The description and the drawings shall, however, be used to interpret the patent claims."

[152] Federal Court of Justice (BGH), Judgment of 12 March 2002 – X ZR 135/01, GRUR 2002, 519 – Schneidmesser II.

[153] Section 14 sentence 1 Patent Act.

Variants of Interpretation of Patent Claims

A legal provision can be interpreted in four ways. There can be a literal interpretation, a systematic interpretation in which "neighboring legal provisions" are also used for understanding, a teleological interpretation in which the purpose intended by the legislator is considered, and a historical interpretation that takes into account the legislative history of the provision.

A patent claim cannot be subjected to a historical interpretation, since the history of the development of an invention cannot be determined and is also not relevant. Historical interpretation is therefore not applicable to a patent claim. A claim can only be interpreted in three ways, and all three variants should be applied in order to reliably determine the scope of protection of a patent. First, the literal meaning of the claim should be determined, which involves asking about the meaning of each individual word of a claim. The totality of the individual meanings results in the literal interpretation of the claim.

Literal Interpretation

In the literal or wording-based interpretation, the sum of the literal meanings of the terms gives the meaning of the claim. The claim is therefore to be understood exactly as it is written. The literal interpretation is the first variant of interpretation to be applied.

Systematic Interpretation

A systematic interpretation is already required by the Patent Act, since the meaning of the claim is to be determined in the context of the entire disclosure of the patent specification.[154] In systematic interpretation, it may in particular be found that terms are to be understood differently than usual. Systematic interpretation can resolve ambiguities and inconsistencies in claims.

The systematic interpretation serves to interpret the independent claims, which define the scope of protection in light of the drawings and the description of the patent. However, this must not lead to an limitation of the scope of protection to the embodiments described by the specification of the patent.

Teleological Interpretation

A teleological interpretation seeks the objective that a claim is intended to fulfill. Therefore, the purpose of the claim is investigated.[155] A teleological interpretation is objective-oriented, as the subject matter of the claim to be interpreted is construed to the extent required by the solution of the technical objective.[156] Only those embodiments are considered to belong to the scope of protection that solve the defined technical objective.

---

[154] Section 14 sentence 2 Patent Act.

[155] *BGH, Judgmentof 4.2.2010 – Xa ZR 36/08, GRUR 2010, 602 – Joint arrangement.*

[156] BGH, Judgment of 4.2.2010 – Xa ZR 36/08, GRUR 2010, 602 – Joint arrangement.

A teleological interpretation can result in a scope of protection that is too narrow, since embodiments that do not fulfill the technical objective may also fall within the scope of protection. Accordingly, the teleological interpretation is more suitable for confirming that already identified embodiments are included in the scope of protection, rather than for excluding individual embodiments from the scope of protection. A particular difficulty of the teleological interpretation is finding the appropriate technical objective.

Person Skilled in the Art

The interpretation of a claim is carried out from the perspective of a person skilled in the art in the relevant technical field. The patent law skilled person is a fictitious average skilled person with average ability and expertise in the pertinent technical field.[157] As a rule, the patent law skilled person has an engineering education and many years of professional experience in the relevant technical field.[158] A skilled person will understand a claim such that only technically meaningful embodiments are encompassed by the scope of protection of the claim.

2.16 How do I Write a Patent Application?

This section presents the structure and procedure for preparing a patent application. Afterward, the reader will be able to assess the quality of a draft patent application and also prepare patent applications themselves at an advanced level.

The Components of an Application

A patent application comprises five components: the introductory part, the drawings, the description of the drawings, the patent claims, and the abstract.[159] The abstract serves solely to inform the public and, in the strict sense, does not constitute a component of the disclosure of the invention.[160] Utility models do not have an abstract, and the claims in a utility model are not called patent claims, but rather protection claims. Otherwise, a utility model application is very similar to a patent application.

[157] BGH, Judgment of 9 January 2018 – X ZR 14/16, GRUR 2018, 390 – Heat energy management; BGH, Judgment of 12 March 2002 – X ZR 43/01, GRUR 2002, 511 – Plastic pipe part.

[158] BGH, Judgment of 9 January 2018 – X ZR 14/16, GRUR 2018, 390 – Heat energy management; BGH, Judgment of 12 March 2002 – X ZR 43/01, GRUR 2002, 511 – Plastic pipe part.

[159] Article 78(1) EPC (European Patent Convention) or Section 34(3) Nos. 3 to 5 Patent Act.

[160] Article 85, first sentence, EPC.

Introductory Part

The introductory part of the description outlines the logical development of the invention by stating the starting point, i.e., the prior art that the inventor was confronted with, and the technological disadvantages of the prior art, from which the technical objective arose. The technical objective led to the invention, which solves this objective.[161]

The introductory part of a patent application is dedicated to the genesis of the invention. The introductory part should lead up to the invention and clarify the purpose of the invention. This section also presents an opportunity to steer the examiner's search at the patent office at the beginning of the grant phase in a direction for which the patent application has already been prepared. Granting of a patent may thus be facilitated.

In the introductory part of the description, the prior art is first explained. The prior art includes all publications before the filing or priority date. The publications may have been made in written or oral form. In addition, technologies may have been made known to the public by public use.[162]

The prior art should be discussed in detail, and in particular, its disadvantages should be identified, which lead to the technical objective. The technical objective should not be formulated simply as "overcoming the disadvantages of the prior art," but should be specifically stated, for example, as "shortening repair times," "extending service life," "avoiding waste," or "reducing production times."

From the introductory part, it can therefore be determined from which prior art the inventor started. The introductory part can therefore be used very well for a teleological interpretation of the invention (purpose of the invention or which technical objective is solved by the invention).

Before drafting a patent application, a search of the prior art should always be conducted first. A search of the prior art can have two objectives. On the one hand, the purpose of a search may be to determine the patentability of an invention. In this case, all publications prior to the filing or priority date are relevant. Another objective is to determine whether a product can be manufactured without infringing third-party intellectual property rights. In the latter case, patents, patent applications, and utility models that are no longer in force can be disregarded.

From the disadvantages of the prior art identified in the search, the technical problem of the invention is to be derived. Care should be taken to ensure that the technical problem does not describe or clearly anticipate any features that belong to the invention. Otherwise, the inventive step would be presented in a diminished manner.

In addition, the claims should be mentioned in the introductory section. It is also customary to provide explanations for each claim, such as definitions of terms, and to explain the respective advantages of the specific embodiments of the independent claims and the dependent claims.

[161] Section 10(2) No. 2 Patent Regulation.

[162] Section 3(1) sentence 2 Patent Act or Article 54(2) EPC.

Drawings

The drawing is the language of the engineer. Accordingly, it is advisable to include drawings with a patent application.

The drawings of a patent application must be of such quality that the content remains clearly recognizable even after multiple copies. Colored illustrations are not permitted. The extensive formal requirements can be found in Annex 2 to § 12 of the German Patent Regulation.

Drawings can convey a large amount of information very quickly, especially regarding the technical field to which the invention pertains. In addition, the details of the invention can already be illustrated, so drawings should by no means be omitted from the description of the invention. A quick understanding of third-party patents can be achieved in particular through the drawings and by reading the claims.

The drawings illustrate specific embodiments of the invention and generally represent special, particularly advantageous configurations of the invention.

Figure 2.1 shows two patent drawings with reference numerals, from which lines or arrows lead to the element that is to be identified by the respective reference numeral. The lines may be curved in design, so that there is no risk of them being mistaken as part of the illustration.

Patent applications typically include multiple drawings, which may also depict the same elements, for example from different perspectives. To maintain clarity, the same elements should be assigned the same reference numerals.

A patent drawing must meet a number of formal requirements, which are set out in Annex 2 to § 12 of the German Patent Regulation. If these requirements are not met, this will be objected to by the Patent Office and can be corrected within a specified period without jeopardizing the early filing date.

Annex 2 to § 12 of the Patent Regulation determines, in particular, the following points:

> *"The drawings must be made on sheets with the following minimum margins: Top margin: 2.5 cm; left margin: 2.5 cm; right margin: 1.5 cm; bottom margin: 1 cm."*[163]

Minimum page margins must therefore be observed.

> *"For the illustration ofthe invention, in addition to views and sectional drawings, perspective views or exploded views may also be used. Cross-sections must be indicated by hatching, which must not impair the legibility ofthe reference numerals and leader lines."*[164]

Sectional drawings should, as is customary, be indicated by hatching. The arrows or lines from reference numerals to the elements to be identified in a drawing can in particular be designed as curved lines, in order to avoid confusion with parts of the drawing.

[163] No. 1 of Annex 2 to § 12 of the Patent Regulation.

[164] No. 3 of Annex 2 to § 12 of the Patent Regulation.

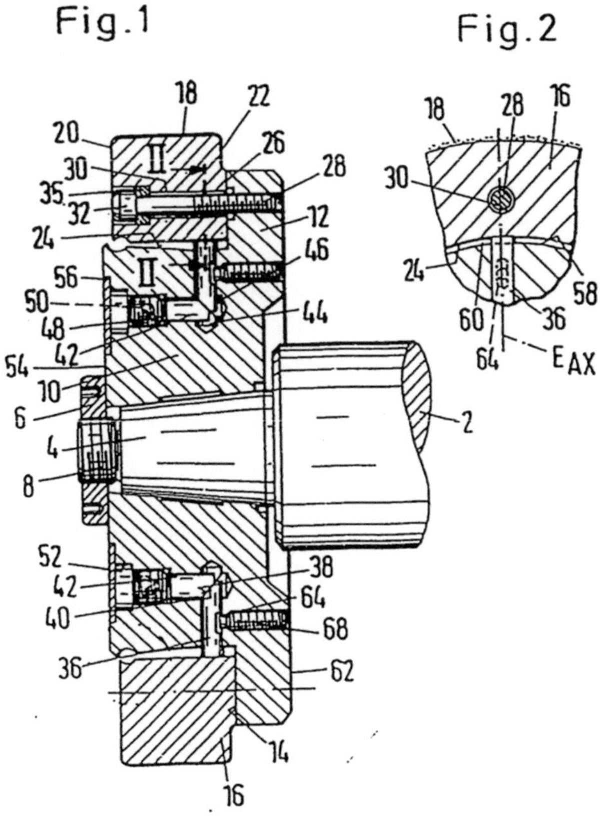

Fig. 2.1 Figs. 1 and 2 of DE3635973 as examples of patent drawings[165]

> *"The lines of the drawings should not be drawn freehand, but with drawing instruments. The numerals and letters used in the drawings must be at least 0.32 cm high. For the labeling of the drawings, Latin and, where customary, Greek letters are to be used."*[166]

The characters should have a minimum height of 0.32 cm to ensure legibility even after multiple copies.

> *"Reference numerals may only be used in the drawings insofar as they are listed in the description and, where applicable, in the claims, and vice versa. The same applies tothe abstract and its drawing."*[167]

It is advisable that reference numerals used in drawings are also used in the description.

---

[165] DPMA, https://depatisnet.dpma.de/DepatisNet/depatisnet?action=pdf&docid=DE000003635973C2&xxxfull=1, accessed on 28.9.2025

[166] No. 5 of Annex 2 to § 12 of the Patent Regulation.

[167] No. 7 of Annex 2 to § 12 of the Patent Regulation.

Description of the drawings

Based on the drawings, the specific embodiments of the invention can be described. In this context, the individual elements of a drawing should first be described, followed by an explanation of the interaction of the elements, which leads to the advantageous effect of the invention.

The description of the drawings comprises two sections, namely a "brief description of the drawings", in which the figures are explained in one or two sentences, and a "detailed description of exemplary embodiments", in which the specific embodiments are described in their individual elements and their interaction.

The description often includes a reference list, in which the individual reference signs are assigned their meaning. A reference list can greatly facilitate an initial understanding of the drawings.

Claims

The claims define the scope of protection of a patent.[168] In this context, the independent claims are particularly relevant. The independent claims do not refer to preceding claims and usually belong to different claim categories. The entirety of the claims of a patent is called the claim set.

There are basically two categories of claims, namely product claims and process claims. With a product claim, a device, in particular a machine or a device for manufacturing a product, or a substance, in particular a chemical composition, can be claimed. A process claim can include steps for manufacturing a product, for applying an analysis method, or for using a chemical substance.

The scope of protection of a process claim includes not only the described process, but also the product that results directly from the process.[169]

A claim can be formulated in a single-part or a two-part form. In a two-part form, the claim is divided into a so-called preamble and a characterizing part. The preamble contains the features that are known from the prior art. The characterizing part particularly includes those features that are responsible for the novelty and the inventive step. In the single-part claim formulation, there is no subdivision of the features. The applicant is responsible for choosing the manner of formulation.[170] Especially in the case of combination inventions, in which only the combination of features leads to the invention and no individual feature can be assigned a particular role with respect to the inventive step, a single-part claim formulation is to be preferred.[171]

---

[168] Section 14 sentence 1 Patent Act or Article 69 paragraph 1 sentence 1 EPC.

[169] Section 9 sentence 2 no. 3 Patent Act.

[170] Section 9 paragraph 1 sentence 1 Patent Ordinance or Rule 43 paragraph 1 EPC.

[171] EPO, https://www.epo.org/law-practice/legal-texts/html/guidelines/d/f_iv_2_2.htm, accessed on 14.8.2022.

The appropriate formulation of claims is the main task in preparing application documents. The broadest possible scope of protection must be sought, since otherwise the patent becomes worthless due to workarounds. The broadest possible scope of protection of a patent is achieved if embodiments that are just still novel and inventive in view of the prior art fall within the scope of protection of the main claim of the patent.

The claim formulations in the application documents represent an initial attempt at formulation. The task of the grant procedure is precisely to find the broadest possible scope of protection by adapting the claim formulations.

A claim is novel if not all features of the claim can be derived from a single publication of the prior art.[172] A claim is based on an inventive step if the subject matter of the claim is not obvious to the person skilled in the art.[173]An invention is obvious if the person skilled in the art would combine the appropriate documents of the prior art and the subject matter of the claim results therefrom.

A claim set in an application for a German patent very often contains ten claims, since for the eleventh and each additional claim, a claim fee of currently €20 is due.[174] A European application regularly comprises fifteen claims, since the sixteenth and each additional claim currently costs €250. From the 51st claim onwards, an additional claim even costs €630.[175]

In the claims, a particular terminology is sometimes used, which has developed through legal practice, in particular case law. In particular, a generalized language is used. For example, with a functional definition (means-plus-function), devices are described that serve a specific purpose. Such formulations include all devices that serve this purpose, without requiring a precise specification of the device. To preempt the objection of the patent office regarding insufficient enablement, at least one specific embodiment corresponding to the functional structure should be included in the description. Functional definitions can be formulated using the purpose indications "for", "to", and "for the purpose of".[176] A device that is therefore generally suitable for the specified purpose is encompassed by a functional definition. However, if there are physical conditions (for example, load-bearing capacity, heat resistance, etc.) that exclude suitability, the corresponding device is not to be assigned to the scope of protection.

The term "consist" should be used with caution, as it is to be understood as exhaustive. Therefore, if the individual features are listed as consisting of the subject matter in

---

[172] Section 3 paragraph 1 sentence 1 Patent Act or Article 54 paragraph 1 EPC.

[173] Section 4 sentence 1 Patent Act or Article 56 sentence 1 EPC.

[174] Fee item 311050 from the annex to Section 2 paragraph 1 (fee schedule).

[175] Article 2 paragraph 1 no. 15 Rules relating to Fees of the EPO.

[176] EPO, Guidelines for Examination, F-IV, 4.13.1, https://www.epo.org/law-practice/legal-texts/html/guidelines/d/f_iv_4_13_1.htm, accessed on 29.8.2022.

a structural definition of an object, even an additional feature would result in the relevant product no longer falling within the scope of protection of the claim. In a structural definition, in contrast to a functional definition, a list of the features belonging to the subject matter of the claim is described. The terms "comprise" and “include” are not exhaustive and are therefore to be preferred in claim drafting.

Non-technical terms in claims are ignored when assessing patentability. For example, a “device for compliance with tax restrictions” is not a technical term.

The word “respectively” should not be used in claims, as it is ambiguous. Depending on the context, “respectively” can be understood as “and”, “or”, “further”, “including”, “also”, “additionally”, “supplementary”, “with”, “besides”, and “depending”.[177]

The terms “approximately” and “about” can be omitted, since a claim is to be interpreted from the perspective of a person skilled in the art, who always takes into account the usual manufacturing tolerances.[178]

The terms “in particular”and “for example” are often found in claims. These terms indicate specific exemplary embodiments and do not limit the scope of protection. However, they prevent an objection of insufficient enablement and can, if necessary, be used to distinguish from the prior art.[179]

The terms “at least” and “at the very least” are rarely found in claims, since the number “a” or “an” also claims a plurality thereof. However, if the claim explicitly refers to 2 or 5, then the claim does not also cover 1 or 3, but only exactly the number 2 or 5.

The terms “maximize,” “optimize,” and “minimize” lead to problem-based formulations and should be avoided in claims as much as possible. Furthermore, indeterminate formulations may arise, since an object can always be further maximized, optimized, or minimized in any state.

Structure of a Patent Application

- Introductory part
  - *Field of the invention*
  - *Backgroundof the invention(prior art)*
  - *Disadvantages of the prior art*
  - *Object of the invention*
  - *Support of the claims in the description*
- Drawings

[177] Schulte/Moufang, Patentgesetz mit EPÜ, 10th edition, § 34 para. 137.

[178] OLG Düsseldorf, judgment of June 28, 2007–2 U 22/06 – Concrete pump; BGH, judgment of November 7, 2000 – X ZR 145/98 – GRUR 2001, 232, 233 – Hole punch.

[179] DPMA, Guidelines for the Examination of Patent Applications, 2.3.3.6, https://www.dpma.de/docs/formulare/patent/p2796.pdf, accessed on 30.8.2022.

- Description of the drawings
  - Brief description of the drawings
  - Detailed description of exemplary embodiments
- Claims
- Abstract

Utility Model

3

Contents

An alternative to patent application is the utility model. The utility model exhibits several particularities compared to a patent or a patent application. However, according to a decision by the highest court, the required inventive step for a legally valid utility model no longer constitutes a difference. Rather, these two types of intellectual property rights differ in terms of the grace period for novelty, the state of the art, and the procedure leading to the respective full protection right.[1]

3.1 Grace Period for Novelty

A significant advantage of a utility model compared to a patent or a patent application is the grace period for novelty. Due to the grace period for utility models, publications by the inventor or his legal successor are not taken into account when assessing novelty and inventive step, provided that these publications occurred no more than six months before

[1] Federal Court of Justice decision of June 20, 2006, X ZB 27/05, - Demonstration Cabinet, https://juris.bundesgerichtshof.de/cgi-bin/rechtsprechung/document.py?Gericht=bgh&Art=en&nr=49839&pos=0&anz=1, accessed on 13.2.2025.

© The Author(s), under exclusive license to Springer-Verlag GmbH, DE, part of Springer Nature 2025

T. H. Meitinger, *Startup Inventor's Handbook*,
https://doi.org/10.1007/978-3-662-72134-6_3

the filing date of the utility model.[2] Accidental publications by the inventor, for example on his own website or through flyers, are therefore not detrimental within the six-month period. If, as a result of such publications, patent protection is no longer possible, utility model protection may still be pursued.

3.2 State of the Art

For a patent, all publications made available to the public before the filing or priority date are considered.[3] In contrast to patents, oral disclosures and public use abroad cannot be cited against utility models.[4] Therefore, the relevant state of the art for utility models is more limited.

3.3 Registration Procedure

Utility model law does not provide for an official examination procedure. Prior to registration, only a formal examination is conducted. Therefore, a utility model is considered an unexamined intellectual property right, in contrast to a granted patent.

[2] Section 3, paragraph 1, sentence 3 of the Utility Model Act.

[3] Section 3, paragraph 1, sentence 2 of the Patent Act or Article 54, paragraph 2 of the EPC.

[4] Section 3, paragraph 1, sentence 2 of the Utility Model Act.

Trademark

4

Contents

A trademark is a cost-effective intellectual property right that can increase significantly in value over the course of business activities. Every startup should therefore protect a trademark, typically choosing between a German trademark and a European Union trademark, which is valid in all EU member states.

4.1 Types of Trademarks

The two main types of trademarks are word marks and figurative marks. Figurative marks are used in particular to protect logos. There is also a hybrid form, the word and design mark. In addition to these common types, there are many exotic forms of trademarks, such as sound marks, color marks, three-dimensional marks, position marks, and thread marks. These exotic forms play no practical role.

© The Author(s), under exclusive license to Springer-Verlag GmbH, DE, part of Springer Nature 2025

T. H. Meitinger, *Startup Inventor's Handbook*,
https://doi.org/10.1007/978-3-662-72134-6_4

Fig. 4.1 Word and design mark "Coca-Cola"

Word Mark

A word mark consists of one or more words, which may be written in any font, size, or color. Internal capitalization is also permitted without affecting the scope of protection. This results in the broad protection afforded by a word mark. For this reason, a word mark can undergo a facelift without requiring a new registration. Often, old and valuable trademarks are word marks. A notable exception is the Coca-Cola script, which is a word and design mark where the text element cannot be altered without risking that the used mark is no longer protected by the registered mark. As a result, the text element could not be adapted, which is why the Coca-Cola script appears outdated. Figure 4.1 shows the word and design mark "Coca-Cola".[1]

Figure 4.2 shows the word mark "Mercedes-Benz".[2]

Figure 4.2 shows the famous word mark "Mercedes-Benz", which is protected by trademark registration in any font, size, and in the usual color schemes.

Figurative Mark

A figurative mark can be used in particular to protect a logo. Figure 4.3 shows the famous bitten apple, for which a figurative mark has been registered.[3]

A figurative mark should preferably be filed in black and white, as this allows the mark to claim all common color combinations. An exception applies if the color scheme is precisely what makes the logo distinctive.

---

[1] EUIPO, https://www.tmdn.org/tmview/#/tmview/detail/US500000073509019, accessed on 30.11.2022.

[2] EUIPO, https://www.tmdn.org/tmview/#/tmview/detail/DE503020150343536, accessed on 1.12.2022.

[3] EUIPO, https://www.tmdn.org/tmview/#/tmview/detail/US500000078156920, accessed on 30.11.2022.

Fig. 4.2 Word mark "Mercedes-Benz"

Fig. 4.3 Figurative mark of Apple Inc.

Word and Design Mark

A word and design mark contains both graphic elements and textual components. The disadvantage, compared to a word mark, is that a facelift of the text components is not possible without risking the loss of trademark protection.

The word and design mark therefore has a limited scope of protection, as even minor changes to the graphic and/or text components can already lead to a loss of trademark protection. Figure 4.4 shows a word and design mark.[4]

A word and design mark is generally not recommended, as the combination of graphic and text elements significantly reduces the scope of protection, since changes to any component can circumvent protection. It is therefore advisable to first assess

[4] EUIPO, https://www.tmdn.org/tmview/#/tmview/detail/CA500000032317000, accessed on 18.10.2024.

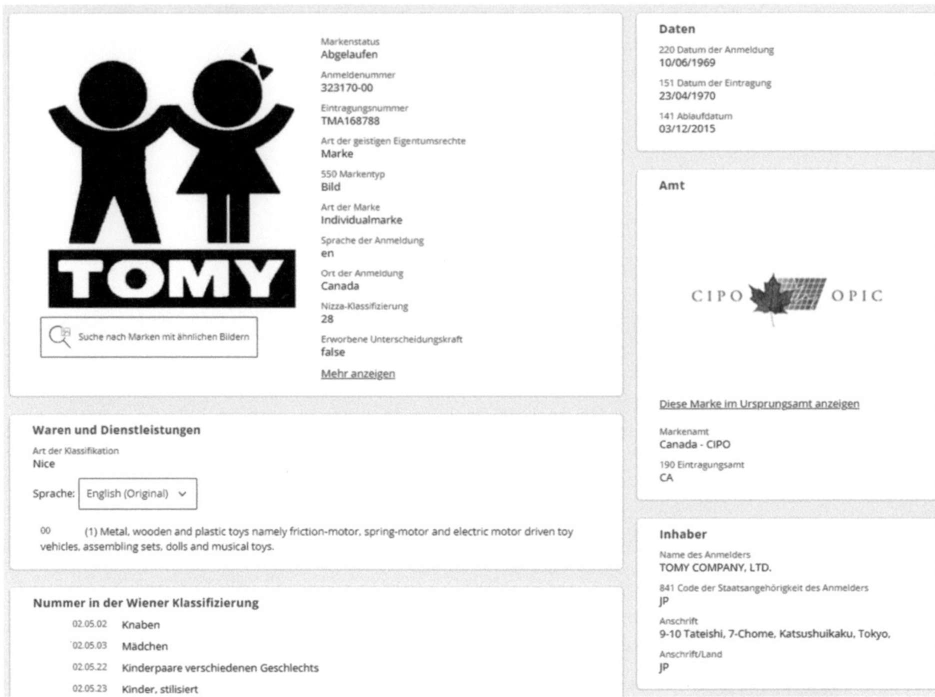

Fig. 4.4 Word and design mark

whether the text components or the graphic elements are worthy of protection. Accordingly, a word mark, a figurative mark, or both can be registered. The less protectable elements can still be used, but they no longer reduce the scope of protection.

Requirement of Use

A trademark must be used; otherwise, it becomes subject to cancellation. The use must be genuine.[5] It is not sufficient for a genuine use to be merely symbolic in order to avoid cancellation.[6] Genuine use is associated with corresponding sales or at least sufficient advertising expenditure for products bearing the trademark. The trademark must be used as it was registered. At the very least, the "distinctive character" must be preserved.[7]

An applicant is not expected to use the trademark immediately after filing or registration. A so-called "grace period for use" is granted, which lasts five years.[8] However, immediately after the grace period expires, genuine use must be established to avoid

[5] Section 26(1) German Trademark Act.

[6] Federal Court of Justice, judgment of June 9, 2011 – I ZR 41/10 – Promotional Gifts.

[7] Section 26(3) German Trademark Act.

[8] Section 49(1) sentence 1 German Trademark Act.

cancellation. The "grace period for use" should therefore be regarded more as a period for commencing use.

The occurrence of cancellation eligibility does not mean that the trademark is immediately deleted by the patent office. Rather, a cancellation action or cancellation proceedings have a prospect of success. If no third party initiates cancellation, the trademark remains in the register. Furthermore, cancellation eligibility that has already occurred can be remedied by commencing use in the meantime. However, this does not apply if cancellation proceedings are already pending.

Descriptive Designation: Still a Trademark?

There is often a desire to register a designation as a trademark that is descriptive of the relevant goods and services. The idea is that a descriptive trademark makes it easier to sell the goods and services. This assumption may be correct, at least for the initial business phase. However, it will become apparent that a descriptive designation will never develop into a valuable trademark. Valuable trademarks are always imaginative. Examples include the trademarks "Amazon", "Google", "Yahoo", and "Apple".

A purely descriptive designation cannot be registered as a trademark due to trademark law.[9] One can attempt to obtain a registrable trademark by adding graphic elements. However, each additional element reduces the scope of protection, so that at best only a trademark of little value will be registered with the patent office.

4.2 Life Cycle of a Trademark

The various stages that a trademark typically goes through are described. These stages are usually: examination of registrability, search for prior rights, trademark application, opposition proceedings, cancellation proceedings, monitoring, and enforcement. The stages "opposition proceedings" and "cancellation proceedings" will not occur if the trademark does not conflict with any prior rights.

Before a trademark is filed with the patent office, it should be checked whether it is fundamentally registrable. An application for a non-registrable trademark will be rejected by the patent office. In addition, a search for prior rights should be conducted to avoid opposition proceedings.

Examination of Registrability

Before a trademark is entered in the register, the patent office examines its fundamental registrability. In particular, the criteria of distinctiveness[10] and the need to keep free[11] are

[9] Section 8(2) no. 2 German Trademark Act.

[10] Section 8(2) no. 1 German Trademark Act.

[11] Section 8(2) no. 2 German Trademark Act.

relevant. A trademark is distinctive if it can be recognized as an indication of origin for the manufacturer or provider of the goods or services. The need to keep free is violated if a trademark application describes the relevant goods or services. This is to ensure that every provider can describe their goods. Monopolizing a description of a product would run counter to this principle. In addition, a trademark must not endanger public order or be contrary to accepted principles of morality. State emblems, national flags, and official certification and guarantee marks are also not registrable.

Search for Prior Rights

The patent office does not conduct an examination for likelihood of confusion with prior rights before entering a trademark into the register. The registration and entry of new trademarks in the trademark register can already lead to market confusion if they collide with older trademarks. This may result in liability for damages. Therefore, an applicant should conduct their own search for prior rights before filing their trademark with the patent office.

When assessing whether an older trademark conflicts with a newer one, it should be remembered that a trademark is a combination of a trademark representation and the associated goods and services. An older right is not automatically relevant simply because the same trademark representation exists. A well-known example is the trademark "duplo". There is a "duplo" trademark for chocolate bars and a "duplo" trademark for interlocking building blocks for children. Since the goods and services for which the trademarks are registered are dissimilar, there is no likelihood of confusion, even though the same trademark representation "duplo" is used. The reason is that market participants do not assume that the same provider manufactures both chocolate bars and toys. In this case, the trademarks can therefore coexist. Figure 4.5 shows the "duplo" trademark for toys.[12] Figure 4.6 shows the "duplo" trademark for chocolate bars.[13]

Filing the Trademark

A trademark can be filed with the patent office in three ways. An application can be submitted by fax, by mail, or online. The application form for filing by fax or mail can be downloaded at the following link: https://www.dpma.de/docs/formulare/marken/w7005.pdf. Online filing is possible for anyone without additional devices or chip cards.[14] An application for an EU trademark can now only be submitted online to the EUIPO[15].

[12] EUIPO, https://www.tmdn.org/tmview/#/tmview/detail/CH501990000003295, accessed on 18.10.2024.

[13] EUIPO, https://tmdn.org/tmview/#/tmview/detail/DE500000000959526, accessed on 28.12.2022.

[14] DPMA, https://direkt.dpma.de/DpmaDirektWebEditoren/w7005/w7005web.xhtml?jftfdi=&jffi=w7005&jfwid=F98CB04C457C267CDDA4146AC9FEA1A2:0, accessed on 28.12.2022.

[15] EU Intellectual Property Office (European Union Intellectual Property Office) based in Alicante, Spain.

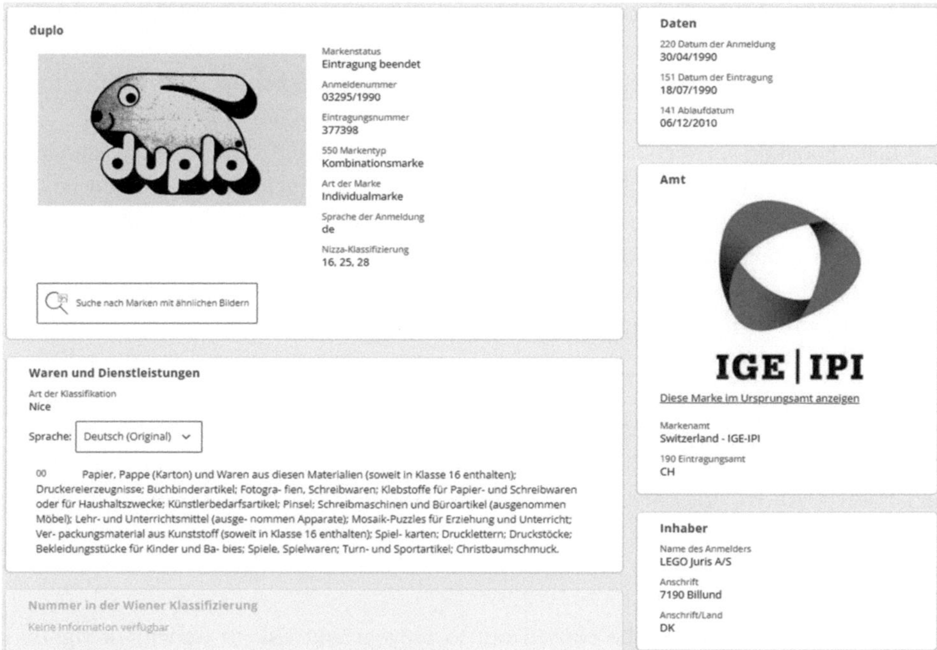

Fig. 4.5 Trademark "Duplo" of Lego Juris A/S

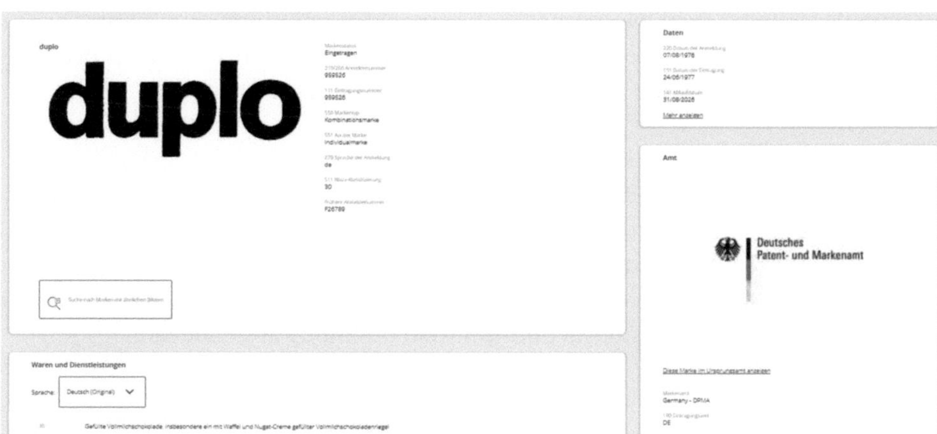

Fig. 4.6 Trademark "Duplo" of Ferrero Deutschland GmbH

Opposition Proceedings

By filing an opposition, the holder of a prior right can have trademarks deleted from the register or prevent their registration in a simple and cost-effective manner if they are likely to be confused with their own trademark. An opposition must be filed within a period of 3 months. In German proceedings, the period begins after publication of the trademark registration.[16] An opposition against an EU trademark can be filed before the trademark is entered in the register. The three-month period begins after publication of the application.

When choosing between a German trademark and an EU trademark, it should be noted that the EUIPO itself conducts a search for older trademarks that could be confused with the newer trademark. The holders of the identified older trademarks are given access to this search and are thus notified of potentially confusing younger trademarks. Therefore, in the European procedure, there is a significantly higher risk that opposition proceedings will be initiated against a trademark.

Cancellation Proceedings

Once the opposition period has expired, a trademark can only be challenged through cancellation proceedings before the regional court[17] or through invalidity proceedings before the DPMA[18]. An EU trademark can be removed through official proceedings before the EUIPO.[19]

Monitoring

Trademark monitoring must be carried out to prevent any risk to trademark rights. If known trademark infringements are not addressed in a timely manner, it may not be possible to take action later. The trademark owner may then have to tolerate inferior goods being offered on the market under their trademark, thereby damaging the brand's image.

Given that a three-month period must be observed for opposition proceedings, a monitoring interval should be chosen that ensures an opposition can be filed in due time. Monthly or at least bimonthly monitoring is therefore advisable.

Enforcement

Trademark rights can be enforced by means of a request for authorization, a warning letter, preliminary injunction proceedings, or litigation. A request for authorization serves to clarify the legal situation, and a warning letter represents the final attempt to resolve a trademark infringement out of court. A preliminary injunction can only be obtained in urgent cases, for example, to stop an infringer from exhibiting at a trade fair.

---

[16] Section 42(1) of the Trademark Act.

[17] Section 55 of the Trademark Act.

[18] Section 53 of the Trademark Act.

[19] Article 63 of the EU Trademark Regulation.

4.3 Grounds for Refusal of Registration

Before a trademark is entered into the trademark register, the Patent Office examines it for so-called absolute grounds for refusal.[20] The most important grounds for refusal are the required distinctiveness of the trademark and compliance with the need to keep certain terms free for general use. In addition, the risk of deception, violation of public order or accepted principles of morality, the use of state emblems, national flags, official certification and guarantee marks, and bad faith in the application can also lead to the rejection of a trademark application. A lack of distinctiveness or violation of the need to keep terms free can be remedied if the trademark has acquired distinctiveness through extensive recognition in the market. The applicant must provide evidence of acquired distinctiveness.

Lack of Distinctiveness

A designation lacks distinctiveness if the relevant public does not recognize it as a trademark at all. For example, words such as "Wow", "Super", "our products are the best", or "First Class" are not eligible for registration, as these words are understood merely as promotional statements.

Need to Keep Free

Terms that are used to describe goods and services cannot be registered as trademarks for those same goods and services. For this reason, the term "bread" is not eligible for trademark protection for bakery products. However, the word "bread" could very well be registrable as a trademark for automobiles.

In particular, signs that describe the type, quality, quantity, intended purpose, value, geographical origin, or the time of production or provision of goods or services are not eligible for trademark registration.[21]

Risk of Deception

If a trademark is likely to mislead as to the nature, quality, or geographical origin of the goods and services, the Patent Office may refuse its registration in the trademark register.[22] A risk of deception exists, for example, if the trademark suggests a Europe-wide presence that does not actually exist.

Public Order and Accepted Principles of Morality

This ground for refusal is not met simply because the trademark is incompatible with a single law. Rather, the sign must be contrary to fundamental principles of the legal

[20] Section 8 of the Trademark Act or Article 7 of the EU Trademark Regulation.

[21] Section 8(2) No. 2 of the Trademark Act or Article 7(1)(c) of the EU Trademark Regulation.

[22] Section 8(2) No. 4 of the Trademark Act or Article 7(1)(g) of the EU Trademark Regulation.

system. The fact that the sign offends the moral sensibilities of a minority is also not decisive.[23]

State Emblems, National Flags, and Official Certification and Guarantee Marks
A sign that contains state emblems, national flags, or other official state symbols is excluded from trademark registration.[24]

Bad Faith
A sign that is applied for solely to hinder a competitor will not be entered into the register.[25] However, if the trademark in question is actually used, there can be no exclusive intent to hinder, and bad faith must be ruled out.

Registration through Acquired Distinctiveness
A trademark may be entered into the trademark register despite a lack of distinctiveness or violation of the need to keep terms free, if it has acquired distinctiveness. Acquired distinctiveness is deemed to exist if the trademark has achieved a nationwide recognition rate of over 50%.[26]

4.4 Likelihood of Confusion

A likelihood of confusion between a later trademark and an earlier trademark gives rise to a right to injunctive relief for the owner of the earlier trademark and may mean that an opposition, cancellation, or invalidity proceeding against the later trademark is successful.[27] A likelihood of confusion exists if the relevant public assumes that the later and the earlier trademark designate goods or services from the same provider or manufacturer.

A likelihood of confusion exists when the signs and the goods and services of two trademarks are similar. If the signs are dissimilar, for example "Mercedes-Benz" and "BMW," there is no likelihood of confusion even if the goods and services are identical. If the signs are similar or even identical, for example the sign "duplo," but the signs are used for dissimilar goods, such as children's toys and chocolate bars, a likelihood of confusion may also not arise. Therefore, a likelihood of confusion is only to be assumed

[23] Section 8(2) No. 5 of the Trademark Regulation or Article 7(1)(f) of the EU Trademark Regulation; Federal Court of Justice, 18.9.1963, Ib ZB 21/62, GRUR, 1964, 136 – Swiss Cheese.

[24] Section 8(2) No. 6 of the Trademark Act.

[25] Section 8(2) No. 14 of the Trademark Act.

[26] Federal Court of Justice, judgment of 29.07.2021, I ZR 139/20, GRUR 2021, 1199 – Goldhase III; Federal Court of Justice, decision of 23.10.2014, I ZB 61/13, GRUR 2015, 581 – Langenscheidt Yellow; Federal Court of Justice, decision of 09.07.2015 – I ZB 65/13, GRUR 2015, 1012 – Nivea Blue.

[27] Section 14(2) of the German Trademark Act or Article 9(2) of the EU Trademark Regulation.

if both the signs and the goods and services for which the signs are used are similar. In cases of double identity, a likelihood of confusion is also assumed. Thus, the similarity of the signs must always be assessed together with the similarity of the goods.

Similarity of signs may arise in (visual) graphic, phonetic, or conceptual form. In the (visual) graphic comparison, the visual appearance of the two trademarks is compared. A phonetic likelihood of confusion arises in the case of phonetic similarity. Two trademarks are conceptually similar if the relevant public associates similar ideas with both signs. An example is the word mark "yellow house" and a figurative mark depicting a yellow house. It is sufficient for a likelihood of confusion if similarity exists in just one of the visual, phonetic, or conceptual comparisons.

In visual and phonetic comparisons, signs that differ significantly in length are considered dissimilar. The beginning and end of the word are more important in the assessment than the middle of the word.

In phonetic comparison, attention should be paid to the sequence of syllables and vowels. A syllable consists of one or more sounds pronounced together. Thus, a text is a sequence of syllables. The vowels a, e, i, o, and u are more significant than consonants. In phonetic comparison, it should also be considered that individual letters may be pronounced similarly. This is especially true for "B" and "P" or "C" and "K."

In visual comparison, it should not be forgotten that a registered word mark claims protection for all fonts, font sizes, and common color schemes.

When assessing the similarity of goods and services, factors such as the type of goods and services, their intended use, whether they are complementary or competing goods, the distribution channels, and the target consumer groups may be considered. For the type of goods and services, aspects such as composition, function, nature, and appearance may be relevant.

An intended use, for example, is personal care, so that shampoo and shower gel are considered similar goods.

Complementary goods are generally considered highly similar. An example is tobacco and cigarette paper. Competing goods are also more likely to be regarded as similar, as the relevant public assumes that providers wish to offer their customers a choice of different products. Competing products include, for example, margarine and butter.

The same distribution channels indicate a similarity of goods, although the common distribution channel "supermarket" is only a weak indicator of similarity, since a supermarket offers a wide variety of goods.

If goods or services are offered to the same consumer groups, this supports a similarity of goods or services.

If, in assessing the similarity of trademarks, the earlier trademark is found to have increased recognition, a likelihood of confusion is more likely to be assumed. If there is no increased recognition, an average level of recognition of the earlier trademark is assumed. If an earlier trademark consists mainly of unimaginative or even descriptive elements, it may be assigned a low level of recognition.

If the likelihood of confusion with a famous trademark is being assessed, dissimilarity of goods and services is irrelevant, as a famous trademark extends to all goods and services.[28] The trademarks "Coca-Cola" or "Mercedes-Benz" are examples of famous trademarks.

4.5 Effects of a Trademark

A registered trademark grants its owner claims for injunctive relief, disclosure, presentation and inspection, damages, destruction, and recall. The enforcement of these claims is limited by statutes of limitation, forfeiture, and exhaustion.

Claim for Injunctive Relief

The owner of a trademark may prohibit any unauthorized party from using their trademark.[29] The claim for injunctive relief arises if there is a risk of first-time or repeated infringement. The risk of repeated infringement exists after the first unauthorized use. The risk of first-time infringement arises only if there is a specific reason. Such a reason may be the announcement of a trademark infringement or the registration of the trademark by the infringer.[30]

If there is a risk of first-time or repeated infringement, this can only be eliminated by submitting a cease-and-desist declaration subject to a contractual penalty.[31] The amount of the contractual penalty must reflect the signatory's seriousness in avoiding future trademark infringements.[32]

Claim for Presentation and Inspection

The claim for presentation and inspection enables the trademark owner to assess the extent of a trademark infringement. This claim can therefore be asserted in advance.[33]

[28] Section 14(2) No. 3 of the German Trademark Act or Article 9(2)(c) of the EU Trademark Regulation.

[29] Section 14(2) German Trademark Act; Article 9(2) EU Trademark Regulation.

[30] Federal Court of Justice (BGH), Judgment of 22 January 2014, I ZR 71/12 – REAL-Chips; BGH, Judgment of 14 January 2010, I ZR 92/08, GRUR 2010, 838 – DDR-Logo; BGH, Judgment of 13 March 2008, I ZR 151/05, GRUR, 2008, 912 – Metrosex.

[31] Higher Regional Court (OLG) Frankfurt, NJOZ, 2007, 4312 – Kollektivmarke Volksbank; BGH, 15 February 2007, I ZR 114/04, GRUR, 2007, 871 – Wagenfeld-Leuchte.

[32] BGH, Judgment of 7 October 1982, I ZR 120/80, GRUR, 1983, 127–128 – Vertragsstrafeversprechen.

[33] Section 19a German Trademark Act.

Claim for Disclosure

The trademark owner may assert a claim for disclosure in order to obtain information, after establishing a trademark infringement, to quantify the damages to be claimed.[34]

Claim for Damages

Damages payable by the infringer can be calculated in three ways. The trademark owner may claim either the actual loss incurred or the lost profits. Alternatively, the infringer's profits may be claimed. The first method is the most commonly used, in which damages are determined according to the license analogy, based on the sales achieved and an assumed license rate. The trademark owner is entitled to choose the method of calculation.[35]

Claim for Destruction and Recall

The trademark owner is entitled to the destruction of products that infringe their trademark. The claim for destruction also extends to devices and equipment primarily intended for the manufacture of trademark-infringing products.[36] However, the claim for destruction is subject to a proportionality assessment, which must also take into account the interests of third parties.[37] In addition, the trademark owner is entitled to a recall claim, allowing them to demand that trademark-infringing products be permanently removed from the distribution network.[38]

Limitation, Forfeiture, and Exhaustion

The claims of the trademark owner are subject to a three-year limitation period under the German Civil Code (BGB).[39] In addition, the claims are limited by forfeiture and exhaustion.

Forfeiture occurs if, after becoming aware of a trademark infringement, the owner fails to take action. In addition, the infringer must have used the third-party trademark in good faith.[40]

---

[34] Section 19 German Trademark Act; BGH, Judgment of 29 September 1994, I ZR 114/84, GRUR, 1995, 50 – Indorektal/Indohexal; BGH, Judgment of 26 November 1987, I ZR 123/85, GRUR, 1988, 307 – Gaby.

[35] BGH, Judgment of 2 February 1995, I ZR 16/93, GRUR, 1995, 349 – Objective Calculation of Damages.

[36] Section 18(1) German Trademark Act.

[37] Section 18(3) sentence 1 German Trademark Act.

[38] Section 18(2) German Trademark Act.

[39] Section 20 sentence 1 German Trademark Act in conjunction with Section 195 BGB.

[40] BGH, Judgment of 31 July 2008, I ZR 171/05, GRUR, 2008, 1104 – Haus & Grund II; BGH, Judgment of 21 July 2005, I ZR 312/02, GRUR, 2006, 56 – BOSS-Club; BGH, Judgment of 6 May 2004, I ZR 223/01, GRUR, 2004, 783 – Neuro-Vibolex/Neuro-Fibraflex.

Exhaustion means that a trademark owner cannot exercise control over a product bearing their trademark more than once. If, for example, a trademarked product has been manufactured by a licensee, i.e., with the permission of the trademark owner, the trademark owner can no longer control the further distribution of that specific product and cannot, for example, prevent its sale to certain purchasers.[41]

4.6 German Trademark

A German trademark acquires its protective effect within the territory of the Federal Republic of Germany upon registration in the register of the German Patent and Trademark Office (DPMA: Deutsches Patent- und Markenamt).[42]

The German Patent Office does not examine a trademark application for likelihood of confusion with prior rights before entering it into the register. Instead, an opposition procedure has been established, which allows holders of earlier rights to take action against younger trademarks that present a likelihood of confusion. The EUIPO notifies holders of earlier rights about trademark applications that could potentially give rise to a likelihood of confusion. The German Patent Office does not issue such notifications. The trademark owner must independently monitor for younger trademarks that could potentially endanger their own mark.

Deadlines and Fees

A trademark application will only be processed by the Patent Office if the application fee is paid within three months of the filing date.[43] Payment of the fee initiates a protection period of 10 years. After the initial 10-year term, a trademark can be renewed an unlimited number of times for further 10-year periods, provided that a renewal fee is paid no later than 6 months before the expiry of each protection period.[44] If the deadline is missed, there is a grace period of 6 months, which can be utilized by paying a late surcharge.[45]

The application fee for an online trademark application is currently €290.[46] The application fee includes three classes for goods and services. The trademark should ideally be registered for three classes, as this provides the applicant with "bargaining chips" in the event of a dispute with the holder of an earlier trademark. Frequently, coexistence of previously competing trademarks can be achieved by partially waiving individual classes.

[41] Section 24(1) German Trademark Act.

[42] Section 4 No. 1 Trademark Act.

[43] Section 6 (1) sentence 2 Patent Costs Act.

[44] Section 47 (2) Trademark Act.

[45] Section 7 (3) sentence 2 Patent Costs Act.

[46] Annex to Section 2 (1) – Schedule of Fees – Fee item 331.000.

Priority

A trademark application establishes a six-month priority period from the filing date. Priority may be claimed for any foreign or regional trademark application, such as a European Union trademark, so that the earlier filing date of the German application benefits the subsequent application.[47]

Earlier Rights

Only earlier rights that are effective in Germany can be asserted against a German trademark. This means that only German trademarks, EU trademarks, and international registrations designating Germany are to be considered. The likelihood of a German trademark being challenged is therefore significantly lower compared to, for example, an EU trademark, which can be attacked by national trademarks from all EU member states.

Withdrawal and Surrender

A trademark owner may withdraw their trademark at any time or surrender individual classes or goods and services.[48] By withdrawing or surrendering individual classes, it may be possible to terminate a pending opposition proceeding or enable the coexistence of competing trademarks.

4.7 European Union Trademark

A European Union trademark (EUTM) grants trademark rights in all EU member states. Since "Brexit" in 2020, an EU trademark no longer applies to the United Kingdom.

Deadlines and Fees

The application fee for an EU trademark is currently €850 and must be paid within one month. The application fee includes only one class. An additional class costs €50, and for a third and each further class, €150 is payable.[49] The initial protection period is 10 years and can be renewed an unlimited number of times for further 10-year periods by payment of a fee.[50] The renewal fee may be paid six months before the expiry of the protection period.[51]

---

[47] Article 4 C (1) Paris Convention.

[48] Section 39 (1) Trademark Act and Section 48 (1) Trademark Act.

[49] EUIPO, https://euipo.europa.eu/ohimportal/de/fees-payable-direct-to-euipo, accessed on 20.03.2023; Article 31 (2) EU Trademark Regulation.

[50] Article 52 EU Trademark Regulation.

[51] Article 53 EU Trademark Regulation; EUIPO, https://euipo.europa.eu/ohimportal/de/renewals, accessed on 20.03.2023.

Priority

An EU trademark can establish priority for a national or regional trademark, or itself claim priority. If priority is to be claimed, the EU trademark must be filed with the EUIPO within 6 months of the first application.[52]

Seniority

Seniority allows an earlier filing date of a national trademark of an EU member state to be transferred to an EU trademark. However, the earlier filing date then applies only to the respective national portion, not to the EU trademark as a whole.[53] In this way, a trademark portfolio can be streamlined. The disadvantage is that the EU trademark can be challenged by earlier rights from all EU member states, which is not the case for a national trademark. Thus, while streamlining the trademark portfolio, the risk of losing trademark rights increases. For this reason, seniority is rarely used.

Observations by Third Parties

With "observations by third parties," anyone may inform the EUIPO of facts that preclude registration.[54] "Observations by third parties" may only relate to absolute grounds for refusal.[55] Earlier trademark rights are not taken into account.

Search Report

A special feature of the EUIPO is the preparation of a search report for holders of earlier trademark rights, who are informed about trademark applications that could give rise to a likelihood of confusion.[56] Holders of earlier trademark rights automatically receive information about potentially confusing trademarks. The owner of the younger trademark may request the search report prepared for their trademark.[57]

Surrender

The trademark owner may at any time surrender their trademark or individual goods and services.[58] This may terminate an opposition or cancellation proceeding before the EUIPO.[59] Surrender is a unilateral declaration of intent, the effectiveness of which does not require the consent of the opposing party in any pending bilateral proceedings.

---

[52] Article 4 C (1) Paris Convention.

[53] Article 39 (1) and Article 40 (1) EU Trademark Regulation.

[54] Article 45 EU Trademark Regulation.

[55] Article 7 EU Trademark Regulation.

[56] Article 43 (7) sentence 1 EU Trademark Regulation.

[57] Article 43 (6) EU Trademark Regulation.

[58] Article 57 (1) and (2) sentence 1 EU Trademark Regulation.

[59] Article 46 and Article 63 EU Trademark Regulation.

4.8 International Registration

An international registration can be used to obtain trademark protection in any number of countries or regions. Currently, trademark protection can be claimed in up to 114 countries in this way. The legal basis for an international trademark application is the Madrid Agreement Concerning the International Registration of Marks (MMA) and the Protocol Relating to the Madrid Agreement (PMMA).

Basic Mark

A prerequisite for international registration is a so-called basic mark, which can be either a German trademark or a European Union trademark. It should be noted that the international registration depends on the existence of the basic mark for five years. If the basic mark is successfully challenged during this period, the international registration is also cancelled. A German trademark is therefore recommended as a basic mark, since it can only be challenged by trademarks from Germany. In addition, a lower fee is charged for a German trademark, as Germany is a full member and the EU has only signed the Protocol.

Procedure for International Registration

An international registration does not grant an international trademark, but rather a bundle of national trademark applications, with the desired target countries needing to be specified at the time of application.[60]

Madrid Agreement (MMA)

The member states of the Madrid Agreement are referred to as "full members". Important full members include: Austria, Belgium, China, France, Germany, Italy, Netherlands, Poland, Portugal, Romania, Russia (Russian Federation), Spain, Switzerland, Ukraine, and Vietnam.[61]

Protocol Relating to the Madrid Agreement (PMMA)

The Protocol Relating to the Madrid Agreement (PMMA) is legally independent from the Madrid Agreement (MMA). Important member states include: Australia, Austria, Belgium, Brazil, Canada, China, Denmark, EU, France, Germany, India, Israel, Italy, Japan, Netherlands, Norway, Philippines, Poland, Portugal, Romania, Russia (Russian

---

[60] WIPO, https://www.wipo.int/madrid/en/how_to/file/basics.html, accessed on 27.03.2023.

[61] WIPO, https://www.wipo.int/export/sites/www/treaties/en/docs/pdf/madrid_marks.pdf, accessed on 27.03.2023.

Federation), Spain, Sweden, Switzerland, Turkey, Ukraine, United Kingdom, USA, and Vietnam.[62]

Deadlines and Fees

The applicant for an international registration must pay a basic fee and a fee that depends on the number of desired Nice classes within one month of the application date. In addition, national fees must be paid for the respective countries.[63] WIPO provides a fee schedule that can be accessed online.[64] The Nice Classification is an internationally recognized classification of goods and services for trademark registration purposes.

4.9 Developing Your Own Trademark

An entrepreneur should secure a trademark with the German Patent Office or with the EUIPO. The trademark should be designed in such a way that it can accompany and support the business activity over the long term. Several aspects need to be considered in this regard. In particular, a word mark is recommended, which should be imaginative and by no means descriptive of the goods and services to be designated.

Trademark as a Combination of Representation and Goods

A trademark should be viewed as a combination of a sign and the goods and services that are identified by the trademark. It is therefore entirely possible for identical signs to coexist under different owners, provided they are used for dissimilar goods and services.

Word Mark Preferred

There are many different types of trademarks, such as color marks, sound marks, 3D marks, position marks, and thread marks. In practice, however, only word marks, figurative marks, and word and design marks are used. A word mark offers broad protection, as every font, font size, and color choice is covered by the scope of protection. This makes it possible, for example, to give the mark a "facelift" by using a modern font, without the "new" mark leaving the scope of protection. In practice, a facelift is typically considered every 10 years, as after about this period the trademark owner may find their mark appears "outdated."

[62] WIPO, https://www.wipo.int/export/sites/www/treaties/en/docs/pdf/madrid_marks.pdf, accessed on 27.03.2023.

[63] WIPO, https://www.wipo.int/madrid/en/how_to/file/fees.html, accessed on 27.03.2023.

[64] WIPO, https://www.wipo.int/wipolex/en/text/586445, accessed on 27.03.2023.

No Word and Design Mark!

A word and design mark should be avoided. If a mark has a distinctive textual component, a word mark should be registered. If there is a distinctive logo, a figurative mark is recommended. A word and design mark contains both a textual component and a graphic element. The textual component of a word and design mark is regarded as a graphic element and therefore cannot be changed without jeopardizing the requirement of genuine use. The many elements of a word and design mark also reduce the scope of protection, as such a mark is generally only infringed if all elements are imitated.

No Descriptive Mark!

No elements that are descriptive of the goods and services to be designated by the trademark should be included in the mark. A descriptive mark will generally never become valuable. In addition, there will be problems with the patent office when registering the mark, and in official or legal disputes, a descriptive mark will be granted only low distinctiveness and thus only a narrow scope of protection.

Design Law

5

Contents

A design right can provide effective protection against imitation of one's own product. In the "patent dispute" between Apple and Samsung, which began with a lawsuit in April 2011, the main issue was a design right that claimed the rounded corners of smartphones.[1] Apple ultimately won the dispute and was awarded substantial damages of 539 million USD.[2]

A design right can also be an important form of protection for a technology-oriented company. At the very least, a design right can provide supplementary legal protection.

A design is not examined for protectability before being entered into the register. The patent office only checks formal registration requirements. A design right is therefore an unexamined intellectual property right.

---

[1] Freakonomics, https://freakonomics.com/2012/08/apple-vs-samsung-who-owns-the-rectangle/, accessed on 13.08.2024.

[2] Der Spiegel, https://www.spiegel.de/wirtschaft/unternehmen/apple-und-samsung-beenden-patent-streit-nach-sieben-jahren-a-1215419.html, accessed on 13.08.2024.

© The Author(s), under exclusive license to Springer-Verlag GmbH, DE, part of Springer Nature 2025

T. H. Meitinger, *Startup Inventor's Handbook*,
https://doi.org/10.1007/978-3-662-72134-6_5

40003553-0001.1

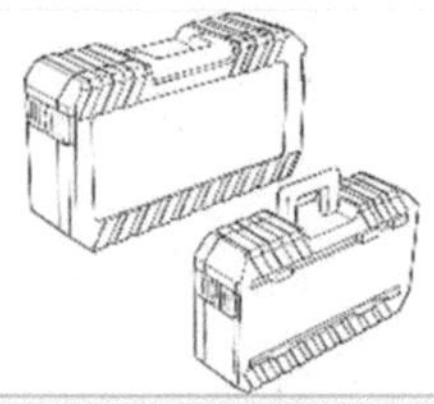

STAMMDATEN

INID	Kriterium	Feld	Inhalt
19	Datenbestand	DB	DE
	Bestandsart	BA	Aktiv
	Designnummer	DNR	40003553-0001
21	Aktenzeichen	AKZ	40003553.7
11	Registernummer	RN	40003553
	Designzustand	DST	Design eingetragen
	Aufschiebungsstatus	AUF	Keine Aufschiebung
22	Anmeldetag	AT	10.04.2000
15	Eintragungstag	ET	07.06.2000
	Veröffentlichungsdatum	VT	10.08.2000
54	Bezeichnung	TI	Werkzeugkoffer
51	Klasse(n)	WKL	03, 08-99
	Klassenversion		7
	Hinterlegungsart		Wiedergabe
	Zahl der Darstellungen	ZD	8
73	Inhaber	INH	Robert Bosch GmbH, 70469 Stuttgart, DE

Fig. 5.1 Tool case (DE40003553-0001.1)

There is no obligation to use a design right. Unlike trademarks, a design right does not become subject to cancellation for non-use.

5.1 Subject Matter of a Design Right

A design right does not require any particular artistic merit. Most registered designs are ordinary craft or industrial products. Figure 5.1 shows, as an example, a tool case from Robert Bosch GmbH.[3] A design can be two- or three-dimensional.

[3] DPMA, https://register.dpma.de/DPMAregister/gsm/register?DNR=40003553-0001, accessed on 13.08.2024.

40004070-0001.1

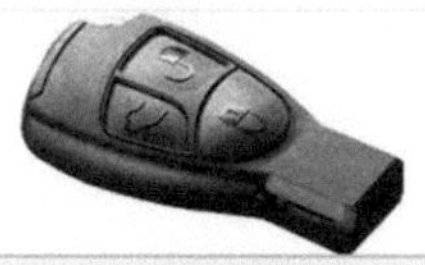

STAMMDATEN

INID	Kriterium	Feld	Inhalt
19	Datenbestand	DB	DE
	Bestandsart	BA	Aktiv
	Designnummer	DNR	40004070-0001
21	Aktenzeichen	AKZ	40004070.0
11	Registernummer	RN	40004070
	Designzustand	DST	Design eingetragen
	Aufschiebungsstatus	AUF	Keine Aufschiebung
22	Anmeldetag	AT	25.04.2000
15	Eintragungstag	ET	07.06.2000
	Veröffentlichungsdatum	VT	10.08.2000
54	Bezeichnung	TI	Schlüssel für Fahrzeuge
51	Klasse(n)	WKL	08-07
	Klassenversion		7
	Hinterlegungsart		Wiedergabe
	Zahl der Darstellungen	ZD	2
73	Inhaber	INH	Mercedes-Benz Group AG, 70372 Stuttgart, DE

Fig. 5.2 Vehicle key (DE40004070-0001)

Figure 5.2 shows a vehicle key from Mercedes-Benz Group AG as another example of a design-protected utility object.

What can be protected as a design right?

A design is the appearance of a product, resulting from the features of lines, contours, colors, shape, surface, or the material from which the design is made.[4] A design can be a craft, industrial, or artistic product.

What can be protected by a design right?

A design does not have to be particularly beautiful or aesthetic to obtain design protection. However, a design must be new and, in particular, possess individual character. Individual character exists if the overall impression of the design is sufficiently distinct from the overall impression of other designs.

[4] Section 1 No. 1 Design Act or Article 3(a) Community Design Regulation.

A disadvantage of design protection is that no general design concept can be protected, but only a specific design, which must be illustrated by representations. The illustrations of the design therefore determine the scope of protection.

Establishment of design protection
A design right is established in particular by the registration of a design application in the register of a patent office. Alternatively, an unregistered Community design can be obtained by publication within the EU. However, such an unregistered Community design is very limited both in time and scope, as it only provides protection for three years and only against identical copying.

Product classes
When registering a design, the products and the associated product classes for which the design is intended to be used must be specified. The indication of products serves solely to classify the design in the register of the patent office. This does not restrict the design. The design owner therefore receives legal protection for all product classes.

Design versus trademark
Aesthetic appearances can be protected by a design right, a figurative mark, or a 3D trademark. In some cases, it may be advisable to protect an aesthetic appearance both with a design right and a registered trademark. The major advantage of trademark law is its potentially unlimited duration of protection, whereas a design right has a maximum term of 25 years.

5.2 Requirements for Design Protection

A design must be visible during normal use and possess individual character in order to constitute a legally valid design right. Special regulations apply to spare parts, must-fit, and must-match components.

Visibility
A design must be visible during its intended use in order to be eligible for design protection.[5] Intended use does not include maintenance or repair of a product incorporating the design, but only use by the end customer. Therefore, a component installed inside a lawnmower engine cannot establish a design right. The same applies to the underside of the lawnmower, as it is not visible during its intended use, i.e., mowing the lawn.

[5] Section 4 Design Act or Article 4(2)(a) Community Design Regulation.

Individual Character and Novelty

A design is considered to have individual character if it produces an overall impression on the informed user that differs from the overall impression of a previously known design.[6] If a design has individual character, it is automatically considered new. For this reason, a separate novelty examination is unnecessary, since if a design exhibits individual character compared to a prior design, it is also new. Conversely, a novelty examination is redundant if the design lacks individual character.

Prior Art

A design right must exhibit individual character over all designs forming part of the prior art in order to be legally valid. The prior art includes all designs that have been made available to the professional circles operating within the European Community.[7] Knowledge of professional circles outside the European Community is irrelevant. Disclosure may occur through publications in newspapers, magazines, books, catalogs, on the internet, at international trade fairs, or on television.

The prior art serves to assess legal validity and to determine the scope of protection, as a design with a high degree of individual character is granted broader protection.

When assessing individual character, the degree of design freedom available to the designer must be taken into account. For example, in the case of scissors, design freedom is limited so as not to compromise the required cutting functionality, meaning that even a small degree of individual character is sufficient to affirm legal validity.[8] Furthermore, design freedom may be limited due to a large body of prior art.

Spare Parts

Spare parts are not eligible for design protection, in particular to limit the influence of automobile manufacturers on the spare parts market.

Must-Fit Component

Must-fit components have a structure that is necessary to achieve the desired interoperability with a complementary element. Such a structure is not eligible for protection, as the legislator intended to prevent an undue extension of protection possibilities.

Must-Match Component

Must-match components are those that not only have a structure to ensure interoperability with a complementary part, but also possess an appropriate appearance to be

[6] Section 2(3) sentence 1 Design Act or Article 6(1) Community Design Regulation.

[7] Section 5 sentence 1 Design Act or Article 7(1) sentence 1 Community Design Regulation.

[8] Section 2(3) sentence 2 Design Act or Article 6(2) Community Design Regulation.

marketable. An example is a car door, which must not only fit into a car body but must also be designed to match the appearance of the rest of the vehicle. Must-match components can not be protected by a design right, because the legislator wants to open up the spare parts business to competition.

5.3 Rights of the Design Holder

A design holder is entitled to injunctive relief to prevent future infringements. Past infringements may be sanctioned by a claim for damages. In addition, there are claims for information, destruction, and recall.

Injunctive Relief

Injunctive relief allows the design holder to prohibit any unauthorized party from using their design. Use is deemed to occur when the design is manufactured, offered, placed on the market, imported, exported, or possessed for these purposes.[9] The right to injunctive relief may be exercised by the design holder and an exclusive licensee. A simple licensee requires authorized legal standing to do so.

Damages

If a design infringement is committed with fault on the part of the infringer, i.e., negligently or intentionally, damages may be claimed.[10] Negligence is already deemed to exist if no search for third-party design rights was conducted, so that, as a rule, fault is assumed.[11]

The design holder may claim lost profits as damages. Alternatively, the surrender of the infringer's profits may be demanded.[12] The design holder may choose between the different methods of calculating damages, with the license analogy method—where a license fee is determined for the calculation of damages—generally being preferred.[13]

Right to Information

The design holder is entitled to information in order to determine the names of suppliers, the quantity of products manufactured, and pricing. This enables the design holder to quantify their claim for damages.[14]

---

[9] Section 38(1) Design Act.

[10] Section 42(2) sentence 1 Design Act.

[11] Federal Court of Justice, 14.1.1958, I ZR 171/56, GRUR 1958, 288, 290 – Dia-Rähmchen.

[12] Section 42(2) sentence 2 Design Act.

[13] Federal Court of Justice, 13.03.1962, I ZR 18/61, GRUR 1962, 401, 404 – Kreuzbodenventilsäcke III.

[14] Section 46(3) Design Act.

Destruction and Recall

Products bearing the protected design must be destroyed if the design holder so requests. The design holder may also demand that infringing products be removed from the distribution chain. In the case of a claim for destruction, proportionality must be observed and the interests of third parties must also be taken into account.[15]

Criminal Provisions

In exceptional cases, intentional design infringement may be punishable by imprisonment.[16]

Limitations of Claims

The claims of the design holder are limited by statutes of limitation, forfeiture, exhaustion, prior rights, and prior use.

For a German design, the three-year limitation period under the German Civil Code (BGB) applies, commencing upon discovery of the infringement.[17] For a Community design valid throughout the EU, the respective national limitation rules must be observed.[18]

Forfeiture applies if enforcement of the claim would violate the principle of good faith.[19] The prerequisite is that the design holder was aware of the infringement for a reasonable period of time, and the infringer acted in good faith and has established a legitimate position.

Exhaustion occurs when the design has been used with the consent of the design holder in a first Member State of the European Union or the European Economic Area. All subsequent uses of the specific product do not require renewed consent from the design holder. This also applies if the first use took place in a Member State of the European Union or the European Economic Area where no design right existed.[20]

An older right to use a design is not restricted or revoked by subsequent design rights.

A right of prior use arising from use of a design before the filing or priority date of a design right is not affected by the design right. The right of prior use is limited to the needs of the business.[21]

---

[15] Section 43 Design Act.

[16] Section 51(2) Design Act.

[17] Section 49 sentence 1 Design Act in conjunction with Section 195 BGB.

[18] Article 88(2) Community Design Regulation.

[19] Section 242 BGB.

[20] Section 48 Design Act.

[21] Section 41(1) sentence 1 Design Act.

5.4 Enforcement of Design Rights

A design right is an unexamined intellectual property right. Therefore, before enforcing a design right, an assessment of its legal validity should be conducted. First, the prior art in terms of existing designs must be identified. The design should then be analyzed using a feature breakdown and assessed for individual character in comparison with the prior art.

Enforcement may initially begin with a request for justification, a warning letter (cease-and-desist letter), or a preliminary injunction. Legal action may also be pursued.

Prior Art in Design

The prior art in design encompasses all designs that were known before the filing or priority date of the design under examination and are therefore relevant to the legal validity of the design.

If there are many similar designs in the prior art, the scope of protection granted to the design may be limited. Conversely, if there is no relevant prior art, the scope of protection for the design is considered to be broad.

Feature Analysis of Overall Impression

In the feature breakdown, the individual features must be specified, enabling a comparison with the designs found in the prior art. Purely technical features are not to be included in the feature breakdown. The individual features should be weighted according to their significance.

Informed User

The overall impression of a design is to be assessed from the perspective of the informed user. The informed user possesses a basic knowledge of designs but is not a design expert. The informed user therefore recognizes details in designs that would remain unnoticed by an ordinary consumer. The informed user is characterized by a high level of attention.[22]

Request for Justification

A request for justification serves to clarify the legal situation, in particular without demanding a cease-and-desist declaration subject to penalty or threatening legal proceedings. The costs of a request for justification are to be borne by the design owner.

Warning Letter

The warning letter is the final attempt to resolve a design infringement out of court. If a warning letter is issued by a patent attorney or a lawyer specializing in intellectual

---

[22] ECJ, 20.10.2011, C-281/10 P, GRUR 2012, 506 – PepsiCo.

property law, it can be assumed that the legal validity of the design has been examined. In such cases, a warning letter should not be dismissed lightly on the grounds that a design is an unexamined right.

The warning letter demands the submission of a cease-and-desist declaration subject to penalty. It also threatens legal enforcement if the infringement is not permanently ceased.

The recipient of the warning letter is usually granted only a short deadline to respond, in order to preserve the option of a preliminary injunction. A response period of one week is reasonable and customary. If an excessively short deadline is set, for example from Friday afternoon to Monday morning, this does not invalidate the warning letter; rather, the short deadline is to be replaced by a reasonable period. The costs of a justified warning letter are to be borne by the recipient; otherwise, by the sender.

Preliminary Injunction

A preliminary injunction allows for a very rapid judicial decision. However, urgency must be present for this to be granted. Therefore, the normal course of legal action must be unreasonable for the design owner.

The problem with a preliminary injunction is that the design right is an unexamined intellectual property right. At the very least, an independent expert opinion on legal validity must be submitted. Nevertheless, the judge involved will very likely not make a decision without giving the opposing party an opportunity to comment. At the very least, a hearing will be scheduled at short notice.

Litigation

Litigation is conducted before an ordinary court, with special regional and higher regional courts designated for design disputes in each federal state to enable specialization in design law.[23]

In litigation, a counterclaim may be filed and the legal validity of the design in question may be challenged.[24] In this case, the court must address not only the infringement but also the legal validity of the design.

Limits of Enforcement

Design law recognizes a right of prior use, which allows anyone who was already using the design before the filing or priority date to continue using it in their business.[25] If a

[23] Section 52 paragraphs 2 and 3 of the Design Act.

[24] Section 52b paragraph 1 sentence 1 of the Design Act.

[25] Federal Court of Justice, 29.06.2017, I ZR 9/16 – Bed Frame, https://juris.bundesgerichtshof.de/cgi-bin/rechtsprechung/document.py?Gericht=bgh&Art=en&az=I%20ZR%209/16&nr=80242, accessed on 06.09.2023.

design is dictated solely by technical function[26] or if the design lacks visibility in normal use[27], enforcement is excluded.

5.5 Invalidity Proceedings

A design can be challenged in invalidity proceedings before the Patent Office or by way of a counterclaim in infringement proceedings.

German Design
A German design can be cancelled through invalidity proceedings before the German Patent and Trade Mark Office (DPMA).

Grounds for Invalidity
A distinction is made between absolute and relative grounds for invalidity. The difference lies in the fact that relative grounds for invalidity can only be asserted by holders of prior rights, whereas absolute grounds for invalidity can be invoked by anyone to initiate invalidity proceedings.

Absolute grounds for invalidity include lack of novelty and lack of individual character of the contested design.[28] If a design is dictated solely by technical function or results from the assembly with a complementary component, it may also be declared invalid. A design is also subject to cancellation if it violates public order or accepted principles of morality.[29]

Relative grounds for invalidity arise from the infringement of prior rights, in particular designs, trademarks, or copyrights. If a design is challenged on the basis of relative grounds for invalidity, it may be possible to remedy the defect and thus prevent the invalidation of the contested design. A remedy occurs if the contested design is modified in such a way that the infringement of the prior right is eliminated. In this context, only the removal of elements from the contested design is permitted. The addition of new elements as a remedy is excluded.

Course of Invalidity Proceedings
The application for invalidity must be substantiated and accompanied by evidence, and filed with the Patent Office.[30] The application is forwarded to the design holder, who is

[26] Section 3 paragraph 1 no. 1 of the Design Act or Article 8 paragraph 1 CDR.
[27] Section 4 of the Design Act or Article 4 paragraph 2(a) CDR.
[28] Section 33(1) No. 2 Design Act.
[29] Section 33(1) No. 3 in conjunction with Section 3(1) Nos. 1 to 3 Design Act.
[30] Section 34a(1) sentences 1 and 2 Design Act.

granted a one-month period to file an objection. If the design holder does not object to the invalidity application, the design is declared invalid.[31]

If the design holder objects, the proceedings become bilateral, and an oral hearing may be scheduled if requested by a party or if the Patent Office considers it expedient.[32]

At the conclusion of the proceedings, a decision on costs may also be made, with the losing party generally bearing all costs of the proceedings (loser pays principle).[33] In the absence of a decision on costs, each party must bear its own costs.[34]

It is possible to join pending invalidity proceedings if a legal interest is demonstrated. Such an interest can be established, in particular, if infringement proceedings concerning the design are pending against the joining party or if the joining party has been warned on the basis of the contested design.[35]

Invalidity proceedings can also be brought against a design that is no longer in force, in order to defend against claims based on the design rights for the past.[36]

Community Design

A design right that has uniform effect in the EU Member States is referred to as a Community design.

A Community design can be challenged by invalidity proceedings before the EUIPO (European Union Intellectual Property Office). In addition, a counterclaim can be filed in infringement proceedings to have the validity of a Community design reviewed.[37]

The application for invalidity is deemed to have been filed only once the invalidity fee has been paid.[38] The application must also be substantiated.

Invalidity proceedings before the EUIPO are conducted exclusively in writing. A design right that is no longer in force may nevertheless be declared invalid for the past.[39]

International Registration

An international registration of a design can be cancelled for the German part by means of proceedings before the German Patent and Trade Mark Office.[40]

---

[31] Section 34a(2) sentences 1 and 2 Design Act.

[32] Section 34a(3) Design Act.

[33] Section 91(1) sentence 1 Code of Civil Procedure.

[34] Section 34a(5) sentence 4 Design Act.

[35] Section 34c(1) Design Act.

[36] Section 33(5) Design Act.

[37] Article 24(1) CDR.

[38] Article 52(2) CDR.

[39] Article 24(2) CDR.

[40] Section 70(1) sentence 1 Design Act.

5.6 German Design

A design belongs to its creator or their legal successor. If the design was created jointly by several designers, the right to the design is held collectively by the group of designers.[41] If the design is created by an employee, ownership is transferred to the employer.[42] This does not apply in the case of a contractual relationship. In such cases, ownership of the design remains with the designer.

If a design is registered by an unauthorized person, the rightful owner may seek cancellation of the registration through a cancellation procedure. Alternatively, the rightful owner may request that the design right be transferred to them.[43]

The applicant is required to name the designer.[44] This is intended to enable the designer to gain recognition. In the case of a group of designers, each designer must be named. The use of pseudonyms does not fulfill the requirement to name the designer.[45]

German design protection is an unexamined intellectual property right, as only a formal examination is conducted prior to registration in the register.[46] There is no examination of protectability, in particular whether the design possesses individual character compared to the prior art.

Application and Registration Procedure

The applicant for a German design application may be a natural or legal person. Joint applications are permitted.

An application must include the following components:[47]

- a request for registration of a design, signed by the applicant,
- the identity of the applicant,
- a representation of the design, and
- an indication of the products for which the design is to be used.

A filing date is accorded to the design application if the first three items are fulfilled.

The following information may also be included with the design application:

- the classes of goods to which the products belong,
- the name of the designer,

---

[41] Section 7(1) Design Act.

[42] Section 7(2) Design Act.

[43] Section 9(1) sentence 1 Design Act.

[44] Section 10 sentence 1 Design Act.

[45] Section 10 sentence 2 Design Act.

[46] Section 16(1) Design Act.

[47] Section 11(2) and (3) Design Act.

- the appointment of a representative,
- an explanation of the representation of the design, and
- a request to defer publication of the representation of the design.

The indication of products and classes of goods does not limit the scope of protection of the design. This information is used solely to classify the design in the register of the Patent Office.[48]

The explanation of the design may be used, in particular, to declare certain elements of the representation as not being part of the claimed design. This can serve to broaden the scope of protection.[49]

Multiple Application

A multiple application allows up to 100 designs to be filed simultaneously, with only one application fee payable. The designs in a multiple application may belong to different classes of goods.

A multiple application may be divided.[50] Before maintenance fees become due, those designs in a multiple application that are no longer needed may therefore be abandoned.

Priority

According to the Paris Convention for the Protection of Industrial Property (Paris Convention), the priority of a first application may be claimed within six months of the filing date of that application.[51] If the priority of an earlier design application is claimed, the filing date, country, file number, and a copy of the earlier design application must be submitted to the Patent Office within 16 months of the filing date of the earlier application.[52]

Deferment of Publication of the Representation of the Design

It is possible to request that the representations of the design be kept confidential for up to 30 months. In this case, only the fact that a design has been registered is publicly accessible. The representations of the design cannot be inspected.[53]

Further Processing

If the applicant misses a deadline for taking an action, resulting in the rejection of the application, the action may be completed within one month. The rejection then becomes

---

[48] Section 11(6) Design Act.

[49] Section 10(2) sentence 1 Design Ordinance.

[50] Section 12(1) Design Regulation.

[51] WIPO, https://www.wipo.int/edocs/pubdocs/de/intproperty/201/wipo_pub_201.pdf, accessed on 31.08.2023, Article 4 C(1) Paris Convention.

[52] Section 14(1) sentence 1 Design Act.

[53] Section 21(1) Design Act.

void.[54] In addition, a fee for further processing must be paid within the one-month period.[55]

Official Fees and Deadlines

An application fee must be paid within three months of filing a design application.[56] Maintenance fees are payable for the 6th to 10th, 11th to 15th, 16th to 20th, and 21st to 25th year. Maintenance fees must be paid in advance, specifically after 5, 10, 15, and 20 years, respectively.[57]

5.7 European Design (Community Design)

A Community design, which provides uniform protection in all EU member states, can be registered with the EUIPO[58] in Alicante. In addition, there is an unregistered Community design, which is established through disclosure.

An application can be submitted to the EUIPO online. Alternatively, the application can be filed with the German Patent Office, which will forward it to the EUIPO.[59] A registration fee and a publication fee are payable. If deferred publication is requested with the application, a deferment fee must be paid instead of the publication fee.[60]

The requirements for a valid Community design correspond to those of the German design.[61] The maximum term of protection, as with the German design, is 25 years.[62]

Right to the Design

The right to the Community design belongs to the designer or their legal successor.[63] However, if the designer is employed and has created the design on the instructions of an employer or in the course of their professional duties, the design is owned by the employer.[64]

[54] Section 17(1) and (2) Design Act.

[55] Section 6(1) sentence 1 Patent Costs Act.

[56] Section 6(1) sentence 2 Patent Costs Act.

[57] DPMA, https://dpma.de/service/gebuehren/designs/index.html, accessed on 31.08.2023.

[58] European Union Intellectual Property Office. The EUIPO was known as the Office for Harmonization in the Internal Market (Trade Marks and Designs), or OHIM, until March 23, 2016.

[59] Article 35 paragraphs 1 and 2 CDR.

[60] EUIPO, https://www.euipo.europa.eu/de/designs/before-applying/fees-payable-direct-to-the-euipo, accessed on 01.09.2023.

[61] Article 4 CDR.

[62] Article 12 sentence 2 CDR.

[63] Article 14 paragraph 1 of the Community Design Regulation (CDR).

[64] Article 14 paragraph 3 CDR.

Deferment of Publication

Well-known companies, such as Apple Inc., Microsoft Corporation, or automotive manufacturers whose new products attract significant media attention, often wish to unveil their products for the first time at specific promotional events. Until then, the new products are to remain confidential. For this reason, it is possible to request deferment of publication. Deferment can be requested for a maximum of 30 months. If priority is claimed for the design, the maximum deferment period is reduced accordingly.[65]

Indication of the Designer

The designer of the Community design must be named.[66] If the design was created by a team of designers, at least the name of the design team must be indicated.[67]

Multiple Application

A multiple application is only possible if the respective products belong to the same class according to the Locarno Classification.[68]

Grace Period for Novelty

Disclosures of the design made by the designer or their legal successor are disregarded when assessing novelty and individual character, provided they occurred no more than 12 months before the filing date.[69]

Unitary Character

A Community design does not constitute a bundle of national rights, but is a single right that has the same effect in all EU member states. If the Community design is transferred, surrendered, or declared invalid, this applies to all EU member states.[70]

Unregistered Community Design

There is an unregistered Community design for the EU area.[71] Instead of registration, disclosure within the European Union is required for the right to arise.[72] The disclosure must result in the design becoming known to the relevant circles operating within the

---

[65] Article 50 paragraph 1 CDR.

[66] Article 18 sentence 1 CDR.

[67] Article 18 sentence 2 CDR.

[68] Article 2 paragraph 1 of the Implementing Regulation to the Community Design Regulation (CDIR).

[69] Article 7 paragraph 2 CDR.

[70] Article 1 paragraph 3 CDR.

[71] Article 1 paragraph 2 lit. a CDR.

[72] Article 11 paragraph 1 CDR.

European Union.[73] The term of protection for the unregistered Community design is limited to three years from the date of disclosure.[74]

In addition to the very short term of protection, the scope of protection is limited. An unregistered Community design can only be used to combat imitations. Furthermore, the infringer must have acted with knowledge of the Community design.[75]

One should not rely on the unregistered Community design. In particular, to enforce it, the intent of the infringer must be proven and the infringement must be an exact copy of the protected design. The unregistered Community design should be regarded as a last resort that can provide assistance in exceptional cases.

5.8 International Design

At WIPO (World Intellectual Property Organization) in Geneva, an international design right can be registered on the basis of the Hague Agreement Concerning the International Registration of Industrial Designs (Hague Agreement).

The international registration of a design is subject to a basic fee and a publication fee. In addition, a designation fee must be paid for each country in which protection is sought.[76] WIPO provides a fee calculator at https://www.wipo.int/hague/en/fees/calculator.jsp.

5.9 Registering a Design

The scope of protection of a design is essentially determined by the representations. Therefore, great importance should be placed on meaningful representations.

Naturalistic or Schematic Representation of the Design

A naturalistic representation provides a detailed reproduction of the design. A naturalistic representation can advantageously be created simply by photographing the design. This allows for a quick and easy generation of the representation. In addition, no important details of the design are omitted. It ensures that the correct overall impression is conveyed. A disadvantage of the naturalistic representation is that the scope of protection may be limited, as the many details depicted can result in a narrower scope of protection.

---

[73] Article 11 paragraph 2 sentence 1 CDR.

[74] Article 11 paragraph 1 CDR.

[75] EUIPO, https://euipo.europa.eu/ohimportal/de/designs-in-the-european-union, accessed on 01.09.2023.

[76] WIPO, https://www.wipo.int/hague/en/, accessed on 01.09.2023.

In a schematic representation of the design, drawings are used. The advantage here is that unimportant details can be omitted, which can result in a broader scope of protection.

Representation in Black and White or in Color

There are designs that are characterized in particular by an unusual use of color. In such cases, the essential part of the overall impression is determined by the color scheme of the design. In this case, a black-and-white representation cannot be submitted. In all other cases, a black-and-white representation should be preferred, as this ensures that all common color variations are covered by the scope of protection.

If a design is registered as a color representation due to its distinctive colors, it is advisable to additionally include a black-and-white representation as a separate design in the application.

Selection of Representations

Typically, a design is shown from the front, back, right, left, top, and bottom. However, submitting all views may be disadvantageous if this reveals details that are not part of the design to be protected, but which could reduce the scope of protection. There is a risk that a modification to an insignificant element could allow circumvention of the design right.

Products and Classes of Goods

The application must specify the products and their classes of goods for which the design is intended to be used. This does not restrict the scope of protection. The Locarno Classification, which can be used to determine the classes of goods, is available at the following link: https://www.dpma.de/recherche/klassifikationen/designs/index.html#a5.

German Design Application

A German application can be filed online or by post with the German Patent and Trade Mark Office (DPMA: Deutsches Patent- und Markenamt). A maximum of ten representations may be attached to each application.[77]

A single application may claim several different designs. In this case, the representations must be marked with two Arabic numerals, where the first digit identifies the design and the second digit the representation. The two digits are separated by a period.[78]

The design must be depicted against a neutral background, and the representation must not contain any text or company logo. Furthermore, only one design may be shown in each representation.[79]

[77] Section 7(1) sentence 2 of the Design Regulation.

[78] Section 7(2) sentence 2 of the Design Regulation.

[79] Section 7(3) sentence 3 of the Design Regulation.

The application may include a description of the representations, which must not exceed 100 words and must be submitted on a separate sheet attached to the application.[80] The description serves to explain the representation, and individual elements may be declared as not forming part of the design.[81] This allows unimportant details in naturalistic representations to be excluded from the determination of the scope of protection, thereby achieving broader protection.

For a surface design, such as wallpaper, at least two views should be submitted.[82] The first view should show the complete design, and the second should show the repeating pattern.[83]

European Design Application

An application can only be filed online. Each design may have a maximum of seven representations.[84] Each representation must be marked with two Arabic numerals, where the first digit identifies the design and the second digit the representation. The two digits are separated by a period.[85]

The design must be shown against a neutral background. The representation must not be edited with correction fluid.[86]

A surface pattern, such as wallpaper, must be shown with at least two representations: one view of the complete pattern and another view showing how the claimed repetition is achieved.[87]

[80] Section 10(2) sentences 1 and 2 of the Design Regulation.

[81] Section 10(1) of the Design Regulation.

[82] Section 8(1) of the Design Regulation.

[83] Section 8(3) of the Design Regulation.

[84] Article 4(2) sentence 1 of the Implementing Regulation to the Community Design Regulation (GGDV).

[85] Article 4(2) sentence 3 GGDV.

[86] Article 4(1)(e) sentence 1 GGDV.

[87] Article 4(3) GGDV.

Employee Invention Law

6

Contents

Employee invention law is an important topic for startups, as its regulations must be observed for technical inventions made by employees.

According to the Patent Act, an invention belongs to the inventor.[1] In contrast, labor law assigns the result of work, including an employee's invention, to the employer. The Employee Inventions Act resolves this contradiction by allowing the employer to claim the invention.[2] In return, the employee is entitled to compensation from the employer.[3]

6.1 Scope of Application

The Employee Inventions Act applies to technical inventions made by employees in both the private and public sectors.[4] The Act governs the rights and obligations of the employer and the inventive employee. The transfer of the invention from the employer

[1] Section 6, sentence 1 of the Patent Act.

[2] Section 6, paragraph 1 of the Employee Inventions Act.

[3] Section 9, paragraph 1 of the Employee Inventions Act.

[4] Section 1 of the Employee Inventions Act.

© The Author(s), under exclusive license to Springer-Verlag GmbH, DE, part of Springer Nature 2025

T. H. Meitinger, *Startup Inventor's Handbook*,
https://doi.org/10.1007/978-3-662-72134-6_6

to a third party does not alter the legal relationship at the time the invention was created. The third party does not become the new "employer" of the inventor under the Employee Inventions Act.

It is important to note that the Employee Inventions Act does not apply to the governing bodies of a company. Therefore, inventions made by a managing director of a GmbH or a board member of a stock corporation are not covered by the Act. If a company wishes to claim the invention of a managing director, this should be stipulated in the employment contract. A clause stating "Inventions of the employed managing director are subject to the Employee Inventions Act" would suffice.

If assignments are given to freelancers, inventions arising from such work also cannot be claimed under the Employee Inventions Act. Therefore, a contractual arrangement should also be made for freelancers.

6.2 Types of Inventions

There are four different types of technical improvements governed by the Employee Inventions Act. A non-patentable improvement is referred to as a technical improvement proposal. If the improvement is patentable, it may be a service invention or a free invention. If a service invention is not claimed or is subsequently released, it becomes a released invention.

Service Invention

A service invention arises in the course of an employment relationship from the employee's professional activity. An invention is also considered a service invention if it is based on the company's know-how.[5] It is irrelevant whether the invention was actually created after working hours, on weekends, or during vacation.

Invention Disclosure

The employee must promptly report an invention to the employer.[6] An invention disclosure must be in written form, which may be provided by letter, email, or fax. The purpose of the disclosure is to enable the employer to decide whether to claim the invention. A proper invention disclosure triggers the deadlines set by the Employee Inventions Act.

The invention disclosure must be submitted separately, and it must be clearly identifiable as a disclosure of an invention. In particular, a disclosure is not considered proper if it is included as one item among several in an email. These formal requirements reflect the significance of the invention disclosure under the Employee Inventions Act. Without a proper disclosure, the statutory deadlines are not triggered.

[5] Section 4, paragraph 2 of the Employee Inventions Act.

[6] Section 5, paragraph 1, sentence 1 of the Employee Inventions Act.

The invention disclosure must explain the technical problem, the fundamental technical concept of the invention, and the various embodiments. In addition, the disclosure must indicate who contributed to the invention and to what extent, which technical resources and company know-how were used, whether the problem and solution were specified by the company, and whether the deficiencies addressed by the invention were identified by the inventor or specified by the company. This information is intended to enable the employer to calculate the employee inventor's compensation.[7]

The employer may request the inventor to provide missing information within a period of two months. If this two-month period expires without such a request, the invention disclosure is deemed proper.[8]

Claiming the Invention

The employer has the option to claim the employee's invention, thereby acquiring ownership of the invention, or to release it. The claim is made by declaration to the employee.[9] If the employer does not wish to acquire the invention, this must be expressly communicated to the employee. Otherwise, a claim is deemed to have been made by law.[10]

If an invention that has already been disclosed is further developed and the development constitutes a patentable invention, the employer must be informed by means of a new invention disclosure. If the further development is not patentable, no disclosure is required, and the development is, in any case, the property of the employer as a result of labor law.

A declaration of claim constitutes a unilateral legal act by way of a declaration of intent, which is effective even against the will of the employee. The claim results in all proprietary rights to the invention being transferred from the employee to the employer. By claiming the invention, the employee acquires a right to compensation from the employer.

Patent Application

The employer must promptly file a patent application for a claimed invention.[11] If there are compelling reasons, the employer may forgo a patent application and use the invention as a company trade secret. The employer is permitted to verify the patentability of the invention through their own search before submitting application documents.

If the employer fails to fulfill the obligation to promptly file a patent application, the employee may file the application at the employer's expense and in the employer's

---

[7] Section 5, paragraph 2 of the Employee Inventions Act.

[8] Section 5, paragraph 3, sentence 1 of the Employee Inventions Act.

[9] Section 6, paragraph 1 of the Employee Inventions Act.

[10] Section 6, paragraph 2 of the Employee Inventions Act.

[11] Section 13, paragraph 1, sentence 1 of the Employee Inventions Act.

name. If the invention loses its patentability due to neglect of the obligation to file promptly, the employer is liable for damages to the employee.

The employee is obliged to cooperate in preparing the patent application documents. Furthermore, the employee must provide the necessary declarations required to obtain the intellectual property right.[12]

Abandonment of Intellectual Property Rights

If the employer has already fulfilled the employee's entire compensation claim, the employer may abandon the patent without notifying the employee. However, if the employee is still entitled to compensation, the employer must offer the patent to the employee for takeover before abandoning the right. The employer must observe the relevant deadlines to ensure that the employee does not suffer a loss of rights upon takeover.

Within a period of three months after the employer has informed the employee of the intention to abandon the intellectual property right, the employee may assert the right to take it over. After the unused period has expired, the employer may abandon the right.[13]

Termination of Rights and Obligations from a Service Invention

When the monopoly position conferred by the intellectual property right ends, the employee's entitlement to compensation also expires. The monopoly position ends in particular if competitors no longer respect the patent, the patent application is rejected by the patent office, or the lack of protectability is established in opposition or nullity proceedings. Compensation payments already made cannot be reclaimed.

Released Invention

If the employer does not claim a service invention or subsequently releases a service invention, this constitutes a "released invention." A service invention is released by means of a written declaration to the employee.[14]

Through this declaration, ownership of the invention reverts to the employee or remains with them. After the release, the employee may dispose of the invention without restriction.[15]

Free Invention

A free invention does not result from professional activity and is not based on the employer's know-how.[16] In contrast to a released invention, a free invention was never a service invention at any time. Nevertheless, a free invention must be reported to the

[12] Section 15, paragraph 2 of the Employee Inventions Act.

[13] Section 16, paragraph 2 of the Employee Inventions Act.

[14] Section 8 sentence 1 Employee Inventions Act.

[15] Section 8 sentence 2 Employee Inventions Act.

[16] Section 4 subsection 3 sentence 1 Employee Inventions Act.

employer so that the employer can verify that it is not a service invention.[17] Notification is not required if it is obvious that the invention cannot be used in the employer's business.[18] As a rule, notification is recommended in order to avoid legal disputes with the employer.

Notification of a free invention triggers a three-month period during which the employer can examine whether it is indeed a free invention. During this period, the employer may contest the status as a free invention. If the employer does not do so, any subsequent claim to the invention by the employer is excluded.[19]

Technical Improvement Proposal

Technical improvement proposals are a special case in industrial property protection, as they are not eligible for protection, yet are addressed by a law on industrial property protection, namely the Employee Inventions Act. A technical improvement proposal must provide the employer with a monopoly-like position in order to give rise to a claim for compensation.[20]

6.3 Entitlement to Compensation

The entitlement to compensation arises from the claimed invention. No intellectual property right is required. The purpose of the compensation claim is to provide the employee with compensation for having to forego the exploitation of their invention. However, the employee should not become a business partner acting on an equal footing with the employer. Rather, the compensation is intended as both recompense and incentive for further inventions.

The amount of compensation regularly leads to conflict between the employee and the employer. To address and defuse this conflict at an early stage, the Official Compensation Guidelines[21] were established and the Arbitration Board[22] was set up at the Patent Office.

Origin and Duration of the Compensation Claim

The compensation claim arises through the claiming of the invention. As soon as an economic benefit or exploitation occurs, the employee is to be given a share of the economic

[17] Section 18 subsection 1 sentence 1 Employee Inventions Act.

[18] Section 18 subsection 3 Employee Inventions Act.

[19] Section 18 subsection 2 Employee Inventions Act.

[20] Section 20 subsection 1 sentence 1 Employee Inventions Act.

[21] DPMA, https://www.dpma.de/docs/dpma/richtlinienfuerdieverguetungvonarbeitnehmererfindungen.pdf, accessed on 22.03.2024.

[22] DPMA Arbitration Board under the Employee Inventions Act, 80297 Munich.

advantage. Economic benefits that do not accrue to the employer are not taken into account. The same applies to economic benefits that are only expected in the future. Losses resulting from the application of the invention are not to be attributed to the employee. Losses do not necessarily exclude compensation.

The obligation to pay compensation exists as long as the employer holds an economic monopoly through the intellectual property right and derives economic benefits from it. If the intellectual property right is no longer respected by competitors or is revoked by an official or judicial procedure, the obligation to pay compensation ends. At the latest, the obligation to pay compensation ends upon expiry of the maximum protection period, unless the employer is still able to maintain a preferential position due to the invention.

Compensation Agreement

The Employee Inventions Act assumes that the employer and the inventive employee will reach an agreement on the compensation for the invention.[23] If the invention was created by a group of inventors, the employer may enter into a separate agreement with each inventor. The agreement may subsequently prove to be inappropriate if circumstances change. In such cases, the agreement may be adjusted. However, repayment of compensation is excluded in all cases.[24]

Determination of Compensation

If the employer and employee cannot reach an amicable agreement, the employer must unilaterally determine the compensation.[25] The determination of compensation must be justified and communicated to the inventive employee in writing, i.e., by email, fax, or letter. The compensation thus determined must be paid by the employer.

The employee may object to the determination of compensation within two months of receipt.[26] The objection must be made in writing. If no objection is raised, the determination of compensation becomes binding for both the employee and the employer.[27]

Adjustment of the Compensation Arrangement

If the circumstances that led to the compensation arrangement change significantly, either the employer or the employee may request an adjustment of the compensation

[23] Section 12 subsection 1 Employee Inventions Act.

[24] Section 12 subsection 6 sentence 2 Employee Inventions Act.

[25] Section 12 subsection 3 sentence 1 Employee Inventions Act.

[26] Section 12 subsection 4 sentence 1 Employee Inventions Act.

[27] Section 12 subsection 4 sentence 2 Employee Inventions Act.

arrangement.[28] The right to request an adjustment may be contractually waived. However, a financial compensation must be agreed upon in such cases.

An adjustment of the compensation arrangement is only possible if it is unreasonable for the contracting party requesting the adjustment to adhere to the existing arrangement. A change resulting from the normal course of business can never justify an adjustment of the compensation arrangement. However, if there is a dramatic economic downturn or the employer is able to serve a market alone because competitors have withdrawn, an adjustment may be requested.

The right to adjust the compensation arrangement is subject to the standard three-year limitation period under the German Civil Code (BGB), with the limitation period beginning when the change in relevant circumstances becomes known. If gross negligence can be attributed to the contracting party, the limitation period begins at that point in time.

Unfairness

A remuneration arrangement, agreement, or determination is invalid if it is already significantly unfair at the outset.[29] Unfairness exists if the remuneration arrangement deviates by more than 50% from the statutory requirements. For high remunerations, a deviation of 25% may suffice. An unfair remuneration arrangement is void from the beginning.[30]

After termination of the employment relationship, unfairness can still be asserted within a six-month period. Text form is sufficient for this purpose, so an email or SMS is enough to meet the deadline. A signature is not required.[31] However, the person making the declaration must be identified.

Non-waivability

Before the creation of an invention, an employee cannot waive rights arising from the invention.[32] Any corresponding clause, for example in an employment contract, is void.[33] However, once an invention is known, the employee may enter into agreements that are disadvantageous to them.[34] Arrangements that are advantageous to the employee are always permissible.

---

[28] Section 12 subsection 6 sentence 1 Employee Inventions Act.

[29] Section 23(1) sentence 1 of the Employee Inventions Act.

[30] Section 139 BGB.

[31] Section 23(2) Employee Inventions Act.

[32] Section 22 sentence 1 Employee Inventions Act.

[33] Section 134 BGB.

[34] Section 22 sentence 2 Employee Inventions Act.

Due date of remuneration

Remuneration becomes due three months after the start of use of the invention.[35] If no use takes place, the remuneration becomes due at the latest three months after the grant of the patent.[36]

Use of the invention

The exploitation of the invention forms the basis for the obligation to pay remuneration. Exploitation can occur through increased sales, license income, or cost savings.

If the question arises as to which invention is to be remunerated, the invention disclosure is decisive, not a patent application or a later patent. Therefore, if an invention is used that is disclosed in the patent but not in the invention disclosure, there is no obligation to pay remuneration. Conversely, the use of an invention that is not claimed in a patent or patent application, but was included in the invention disclosure, must be remunerated.[37]

Experiments and the construction of prototypes intended to achieve market readiness do not constitute use subject to remuneration. The same applies to test sales and trade fair presentations aimed at assessing market acceptance.

Relevant use can also be indirect. For example, if sales are generated as a result of the invention without the invention itself being implemented, there is still an obligation to pay remuneration. "Use" can also arise from a blocking patent if competitors are kept out of a market as a result.

Provision of information

With the commencement of use, the employee is entitled to information from the employer regarding the types and respective extent of use.[38] The right to information does not expire upon termination of the employment relationship.[39]

The right to information also covers foreign uses, regardless of whether employee invention rights exist in the respective country.

The employer is not obliged to provide information about profits or customer and supplier relationships. It is sufficient to state the sales generated, cost savings, and license income for the respective types of use.[40] If the claim for remuneration is only minor, the employer may dispense with comprehensive disclosure.[41] Information should be provided at least once a year.

---

[35] BGH, 28.06.1962 – I ZR 28/61 – "Cromegal", GRUR 1963, 135, 137; BGH, 10.09.2002 – X ZR 199/01 – "Ozon", GRUR 2003, 237.

[36] Section 12(3) sentence 2 Employee Inventions Act.

[37] BGH, 29.11.1988 – X ZR 63/87 "Heavy Metal Oxidation Catalyst" – GRUR, 1989, 205, 207.

[38] BGH, 06.02.2002, X-ZR 215/00 "Wire Injection Device" – GRUR, 2002, 609, 611.

[39] Section 26 Employee Inventions Act.

[40] BGH, 17.11.2009 – X ZR 137/07 "Door Inner Reinforcement" – GRUR, 2010, 223.

[41] BGH, 13.11.1997 – X ZR 132/95 "Copolyester II" – GRUR, 1998, 689, 694.

Calculation of the remuneration claim

If the invention has been filed for patent but the patent has not yet been granted, generally only 50% of the remuneration is to be paid. Once the patent is granted, the withheld portion must be paid in full.

The calculation of remuneration is based on the value of the invention. The value of the invention represents the economic value derived from the use of the invention for the employer. To determine the value of the invention, the license analogy is generally used, whereby the value of the invention is calculated by multiplying the revenue generated by a royalty rate.[42]

Royalty rate

To calculate the value of the invention, a royalty rate is assumed that a reasonable external licensor would demand from an exclusive licensee.[43] The royalty rate will generally range between 0.5% and 1.5%. In the automotive supply industry, a royalty rate significantly below 0.5% may be justified. This also applies to mass-produced items.

Maximum royalty cap

It may occur that several inventions are realized in a single product. In such cases, simply adding up several royalty rates would not be appropriate. In this situation, a maximum royalty rate should be assumed, corresponding to the upper end of the range customary for royalty rates in this technical sector.

Reference base

Typically, a marketable product does not represent the realization of an invention; rather, only a part of the product derives its "distinctive character" from the invention.[44] This part is referred to as the reference base. The revenue attributable to the reference base forms the basis for calculating the value of the invention.

Graduation

Very high revenues are to be graduated, meaning that not the full revenue of the reference base, but a reduced portion is used. Graduation begins at revenues of €1.5 million. The graduation is carried out in steps, initially with a reduction of 10%. From revenues of €2.5 million, a deduction of 20% is appropriate. The exact values can be found in Guideline No. 11 of the "Guidelines for the Remuneration of Employee Inventions in the Private Sector"[45]

---

[42] BGH, 21.12.2005 – X ZR 165/04 "Cylinder Tube", GRUR 2006, 401.

[43] BGH, 13.11.1997 – X ZR 6/96 "Spool Head", GRUR 1998, 684, 687.

[44] BGH, 13.03.1962 – I ZR 18/61 "Cross-bottom Valve Sacks III", GRUR 1962, 401, 402 f.

[45] DPMA, https://www.dpma.de/docs/dpma/richtlinienfuerdieverguetungvonarbeitnehmererfindungen.pdf, accessed on 22.03.2024.

Calculation According to the License Analogy

The value of the invention is calculated according to the license analogy as follows:

- Invention value = $\text{Revenue}_{\text{Reference base}}$ (possibly tiered) · Royalty rate

The remuneration of the inventive employee is determined by multiplying by a participation factor, which takes into account that the employee is not an external licensor:

- Remuneration = Invention value · Participation factor, and thus:
- Remuneration = $\text{Revenue}_{\text{Reference base}}$ · Royalty rate · Participation factor

Participation Factor

The participation factor reflects that the inventive employee benefited from the support of the company, which would not be available to an external licensor. The employee was able to build on the company's know-how and use the employer's machinery, for example, for prototype production.

The participation factor also takes into account that the development of technical inventions is to be expected from a development engineer.

As a rule, the participation factor will range between 10% and 25%, with a typical assignment of 10% to 17% for a development engineer. A senior employee in an R&D department will generally have to accept a participation factor between 5% and 10%.

The calculation of the participation factor is explained in the "Guidelines for the Remuneration of Employee Inventions in the Private Sector"[46], in which three value numbers a, b, and c are determined. These value numbers take into account the circumstances of the task, the company's support, and the inventor's position within the company. Based on a table, the sum of the value numbers a, b, and c can be assigned a participation factor.

Value Number a – Nature of the Task

Value number a reflects the extent to which the inventor was supported by the company in defining the task. In Guideline 31 for the Remuneration of Employee Inventions in the Private Sector[47], the relevant criteria are listed, with the respective value number a indicated in parentheses at the end of each criterion's description:

"The employee was prompted to make the invention:

1. *because the company assigned him a task with direct specification of the solution path taken (1);*

[46] DPMA, https://www.dpma.de/docs/dpma/richtlinienfuerdieverguetungvonarbeitnehmererfindungen.pdf, accessed on 24.03.2024.

[47] DPMA, https://www.dpma.de/docs/dpma/richtlinienfuerdieverguetungvonarbeitnehmererfindungen.pdf, accessed on 24.03.2024.

2. *because the company assigned him a task without direct specification of the solution path taken (2);*
3. *without the company assigning him a task, but through knowledge of deficiencies and needs acquired as a result of company affiliation, if the inventor did not identify these deficiencies and needs himself (3);*
4. *without the company assigning him a task, but through knowledge of deficiencies and needs acquired as a result of company affiliation, if the inventor identified these deficiencies and needs himself (4);*
5. *because he set himself a task within his area of responsibility (5);*
6. *because he set himself a task outside his area of responsibility (6).*"[48]

Regarding 2: As a rule, this category applies to a development engineer, to whom value number a = 2 should therefore be assigned.

Regarding 3: If a development engineer had to overcome internal company resistance to arrive at the invention, value number a = 3 may be appropriate.

Value Number b – Solution of the Task

Value number b reflects the extent of company support provided in the creation of the invention. Guideline 32 defines the parameters for value number b:

"When determining the value numbers for the solution of the task, the following aspects must be considered:

1. *The solution is found using considerations that are professionally familiar to the inventor;*
2. *it is found on the basis of company work or knowledge*
3. *the company supports the inventor with technical resources.*

If all of these criteria are met in an invention, the invention receives value number 1 for the solution of the task; if none of these criteria are met, it receives value number 6. If an invention partially fulfills the three criteria listed, it is assigned a value number between 1 and 6 for the solution of the task."[49]

"Professionally familiar considerations" under the Employee Inventions Act should not be confused with "obvious" as defined in Section 4, Sentence 1 of the Patent Act. An invention may be based on "professionally familiar considerations" and still not be obvious under the Patent Act, and thus be patentable.

---

[48] DPMA, Guideline 31, https://www.dpma.de/docs/dpma/richtlinienfuerdieverguetungvonarbeitnehmererfindungen.pdf, accessed on 24.03.2024.

[49] DPMA, Guideline 32, https://www.dpma.de/docs/dpma/richtlinienfuerdieverguetungvonarbeitnehmererfindungen.pdf, accessed on 24.03.2024.

"Professionally familiar" under the Employee Inventions Act refers to inventions that can be assigned to the technical field in which the employee has worked or been trained for many years. As a rule, criterion no. 1 will therefore be fulfilled. The same generally applies to criteria no. 2 and no. 3, so that value number b is usually assigned the value 1.

Value Number c – Tasks and Position of the Employee in the Company

Guideline 34 takes into account the inventor's position within the company. The corresponding value number is indicated in parentheses at the end of each criterion:

"The following groups of employees can be distinguished, with the value number increasing as the expected level of performance decreases:

Group 8: This group includes employees who essentially have no prior training for the work performed in the company (e.g., unskilled workers, assistants, semi-skilled workers, apprentices) (8).

Group 7: This group includes employees who have received vocational or technical training (e.g., skilled workers, laboratory technicians, fitters, junior draftsmen), even if they already have minor supervisory duties (e.g., foremen, junior supervisors, shift supervisors, team leaders). These individuals are generally expected to perform their assigned tasks with a certain level of technical understanding. On the other hand, it should be noted that this occupational group is generally not expected to solve constructive or procedural technical problems (7).

Group 6: This group includes individuals who serve as lower-level company managers (e.g., supervisors, senior supervisors, plant supervisors) or have received somewhat more advanced technical training (e.g., chemical technicians, technicians). These employees are generally expected to make suggestions for rationalization within their area of responsibility and to be attentive to simple technical innovations (6).

Group 5: This group includes employees who have received advanced technical training, whether at universities or technical colleges, higher technical institutes, or engineering or equivalent technical schools, provided they work in production. These employees are expected to have a keen technical interest and the ability to solve certain constructive or procedural problems (5).

Group 4: This group includes those in production who hold managerial positions (group leaders, i.e., engineers and chemists supervising other engineers or chemists) and engineers and chemists working in development (4).

Group 3: This group includes, in production, the head of an entire production group (e.g., technical department heads and plant managers), in development, the group leaders of design offices and development laboratories, and in research, engineers and chemists (3).

Group 2: This group includes the heads of development departments as well as group leaders in research (2).

Group 1: The top group includes the heads of the entire research department of a company and the technical directors of large enterprises (1)."[50]

Calculation of the Share Factor

The value numbers a, b, and c are added together. The corresponding share factor A can be determined from the following table based on the sum of the value numbers:[51]

a + b + c	=	3	4	5	6	7	8	9	10	11	12	13	14	15	16	17	18	19	(20)
A	=	2	4	7	10	13	15	18	21	25	32	39	47	55	63	72	81	90	(100)

Special Cases of Invention Value

In the vast majority of cases, the value of the invention is determined using the license analogy.[52] An alternative calculation of the invention value is based on the assessment of the company benefit, which is to be applied in cases of exclusive internal company use. If a mathematical determination of the invention value is not possible, only an estimate remains. In addition, there are special cases that allow for an alternative determination of the invention value.

Calculation of the Measurable Operational Benefit

An invention may be used exclusively within the company and therefore may not generate any measurable revenue for the business. In such cases, the license analogy method is not applicable. Operational benefit can arise in particular from the optimization of a manufacturing process, the avoidance of waste or scrap, or the automation of manual tasks. The calculation of the invention's value is based on a comparison of revenues and costs in a scenario with and without the use of the invention.

In practice, it is difficult to determine the respective costs and revenues precisely. Therefore, assumptions and estimates are often necessary, which can compromise the reliability of the method.

As a rule of thumb, 20% of the calculated amount is considered the value of the invention, since the employer's contributions must also be taken into account.

[50] DPMA, Guideline 34, https://www.dpma.de/docs/dpma/richtlinienfuerdieverguetungvonarbeitnehmererfindungen.pdf, accessed on 24.03.2024.

[51] DPMA, Guideline 37, https://www.dpma.de/docs/dpma/richtlinienfuerdieverguetungvonarbeitnehmererfindungen.pdf, accessed on 24.03.2024.

[52] BGH, 21.12.2005 – X ZR 165/04 "Cylinder Tube", GRUR 2006, 401.

Estimating the Value of the Invention

There are inventions whose value can only be determined by estimation. Examples include inventions dedicated to occupational safety or quality assurance. As a rule of thumb, 20% of the estimated benefit is considered the value of the invention, since the employer's efforts must be taken into account.

Value of the Invention in the Case of License Income

License income can serve as the basis for calculating the value of the invention, with the costs required to generate the license income being deducted. In particular, the costs of defending and maintaining a patent must be considered. The resulting benefit can be set at 20% as a rule of thumb for the value of the invention.

Value of the Invention in the Event of a Sale of the Invention

Proceeds from a sale, after deduction of the necessary costs, can serve as the basis for calculating the value of the invention. Typically, a share of 20% to 40% of the calculated benefit is considered an appropriate value for the invention.

Value of the Invention for a Non-Exploited Invention

An employer is not obliged to use an invention. However, this must not result in disadvantages for the inventive employee, so that at the latest after the grant of the patent, the invention must be remunerated.

Value of the Invention for Blocking Patents

A blocking patent is a patent that is not used itself, but serves to secure the revenue of another product. A blocking patent prevents direct competition. To calculate the value of the invention, a share of the secured revenue of the other product can therefore be used. However, only a small portion of the secured revenue can be used for the calculation of remuneration, since the subject matter of the blocking patent is not actually realized.

Value of the Invention for a Utility Model

The value of the invention is determined analogously to the case of a patented invention. However, a deduction of 30% to 50% is appropriate, since a utility model is an unexamined intellectual property right.

Value of the Invention for Foreign Use

Foreign use must be remunerated in the same way as domestic use, even if there is no employee invention law in the relevant foreign country.

Value of the Invention for Trade Secret Inventions

In justified cases, the employer may choose not to file a patent application for the invention and instead use the invention as a trade secret. This does not affect the obligation to pay remuneration or the amount thereof.

6.4 Disputes

The legislator was aware that, in particular, there could be differing opinions regarding remuneration, which could lead to a strain on industrial relations. For this reason, the Arbitration Board[53] was established at the Patent Office, which deals exclusively with disputes arising under the Employee Inventions Act.[54] If an amicable solution cannot be reached with the help of the Arbitration Board, the parties have no option but to pursue legal action.[55]

Arbitration Board

The parties to the employment relationship may refer the matter to the Arbitration Board at any time.[56] A request must be submitted to the Arbitration Board, explaining the facts of the case and specifying the question to be resolved. In addition, the names and addresses of the parties involved must be provided.[57] The request is forwarded to the other parties, who may submit their comments within a period set by the Arbitration Board.[58]

Before legal proceedings can be initiated, the Arbitration Board must be called upon.[59] This does not apply if the employment relationship has already ended[60] or if rights arising from a remuneration agreement[61] or from a settlement proposal of the Arbitration Board are being asserted[62].

It is also not necessary to refer the matter to the Arbitration Board, and legal proceedings may be initiated directly, if it is claimed that an agreement under Sections 12, 19, 22, or 34 is not legally valid[63], if six months have already passed since the Arbitration Board was called upon[64], or if, after a dispute has arisen, the parties have agreed not to make use of arbitration proceedings[65].

After the end of an employment relationship, proceedings before the Arbitration Board are no longer mandatory. However, the Arbitration Board may still be called upon.

---

[53] Section 29(1) Employee Inventions Act.

[54] Section 28 sentence 1 Employee Inventions Act.

[55] Section 37(1) Employee Inventions Act.

[56] Section 28 sentence 1 Employee Inventions Act.

[57] Section 31(1) sentences 1 and 3 Employee Inventions Act.

[58] Section 31(2) Employee Inventions Act.

[59] Section 37(1) Employee Inventions Act.

[60] Section 37(2) no. 3 Employee Inventions Act.

[61] Section 37(2) no. 1 in conjunction with Section 12 Employee Inventions Act.

[62] Section 37(2) no. 1 in conjunction with Section 34 Employee Inventions Act.

[63] Section 37(2) no. 1 Employee Inventions Act.

[64] Section 37(2) no. 2 Employee Inventions Act.

[65] Section 37(2) no. 4 sentences 1 and 2 Employee Inventions Act.

The task of the Arbitration Board is to achieve an amicable settlement between the parties to the employment relationship.[66] To this end, it prepares a settlement proposal[67], which it forwards to the parties involved.[68]

The settlement proposal is not a judicial decision. The parties may file an objection within one month of receipt. If no objection is raised within the respective one-month period, the settlement proposal becomes legally binding on the parties.[69]

Court Proceedings

If arbitration fails, the final recourse is legal action before a civil court. Jurisdiction within a federal state lies with patent litigation chambers at selected regional courts.[70] There are no territorial restrictions for disputes concerning the modalities of payment of a remuneration claim.[71]

---

[66] Section 28 sentence 2 Employee Inventions Act.

[67] Section 34(2) sentence 1 Employee Inventions Act.

[68] Section 34(2) sentence 4 Employee Inventions Act.

[69] Section 34(3) Employee Inventions Act.

[70] Section 39(1) sentence 1 Employee Inventions Act.

[71] Section 39(2) Employee Inventions Act.

Developing a Technology

7

Contents

Based on the state of the art and by taking into account the restrictions of patent law, particularly third-party patents, it is possible to develop or further refine one's own technology to make it market-ready. The state of the art—that is, patents, patent applications, and utility models held by third parties—can be researched using the extensive database "DEPATISnet.dpma.de" provided by the German Patent Office.

As a first step, clarity about one's own technology should be achieved by breaking the technology down into its individual features. This leads to a profound understanding of the technology, which serves as a solid starting point for researching the relevant state of the art. By critically engaging with the state of the art, one's own technology can then be further developed and potentially made patentable.

7.1 Feature Analysis of the Invention

The first step in founding a startup is to analyze your own technology. It is advisable to list and label the individual features of the technology in order to analyze it in detail and compare it with the state of the art.

For example, the idea for a startup might be to create an outdoor play facility for children, with water jets arranged in the ground. This results in a walkable play area from

© The Author(s), under exclusive license to Springer-Verlag GmbH, DE, part of Springer Nature 2025
T. H. Meitinger, *Startup Inventor's Handbook*,
https://doi.org/10.1007/978-3-662-72134-6_7

which water fountains are ejected from the ground—either continuously, alternately, or in a completely unpredictable manner. This also enables games in which the water fountains can be used as barriers for a maze, for instance. Additionally, the water fountains can be illuminated to increase the attractiveness of the facility. The system should be quick to assemble and disassemble to ensure high mobility.

The features of the product are therefore as follows:

- Feature 1: System with a walkable floor with
- Feature 2: water jets, each generating an upward-directed water fountain, where
- Feature 3: the water jets can be controlled individually, where
- Feature 4: lighting elements are arranged on the floor, where
- Feature 5: the lighting elements illuminate the water fountains in different colors, and
- Feature 6: the system can be quickly assembled and disassembled.

This listing of individual features constitutes a detailed analysis of the technology and enables a well-founded comparison with patents, utility models, and patent applications representing the state of the art. Such a listing is referred to as a feature analysis.

7.2 The State of the Art as a Sparring Partner

Researching the state of the art—that is, existing technologies—can serve to assess whether your own technology infringes on third-party patents or is patentable. In particular, the state of the art can be used to further develop your own technology. In this sense, your own technology can be seen merely as a starting point, which can be refined, improved, or even reoriented through "confrontation" with the state of the art, thereby making it market-ready.

The first step is to understand the state of the art and to identify its disadvantages or shortcomings, or to recognize which problems the respective inventors sought to solve and therefore considered important. From this analysis, a new technical objective may emerge, which can be used to further develop your own technology. Of course, this also leads to a better understanding of the market.

The state of the art can present a wide range of technical challenges, allowing for the development of various specialized embodiments that may be advantageous depending on the application or market segment.

Researching the state of the art can be carried out in various ways. A Google search can serve as a starting point. However, one should not stop there. The database of the German Patent Office, "DEPATISnet.dpma.de," offers a much more comprehensive wealth of technical teachings, which are advantageously described in detail in the searchable patent documents.

Fig. 1

Fig. 7.1 Fig. 1 of US20030073505A1

Disadvantages of the State of the Art

Once the relevant state of the art has been identified, it should be examined in detail, with particular attention paid to its shortcomings and disadvantages, which can be used as opportunities to create a better technology.

For example, in the case of the outdoor play facility for children, the document US 20030073505A1 can be found in the state of the art, which discloses an "interactive play fountain."

The main claim of the document describes a water park amusement device with a walkable surface that has openings from which water fountains are ejected by water jets. The openings are shaped and sized so that they can be closed by a hand or foot, but are small enough to prevent a child's foot from slipping inside. If an opening is closed by a user, the water is redirected to another opening.[1]

Figure 7.1 shows a small hill as the ground surface, from which water fountains are ejected.

Figure 7.2 shows the functional principle of the play device, in which an opening 32 is closed by a foot 40, causing the water to be redirected to adjacent openings 32. Nozzles 30b, 30c, and 30d are arranged in the openings 32 to focus the water jet.

[1] DPMA, https://depatisnet.dpma.de/DepatisNet/depatisnet?action=pdf&docid=US020030073505A1&xxxfull=1, accessed on 7.8.2024.

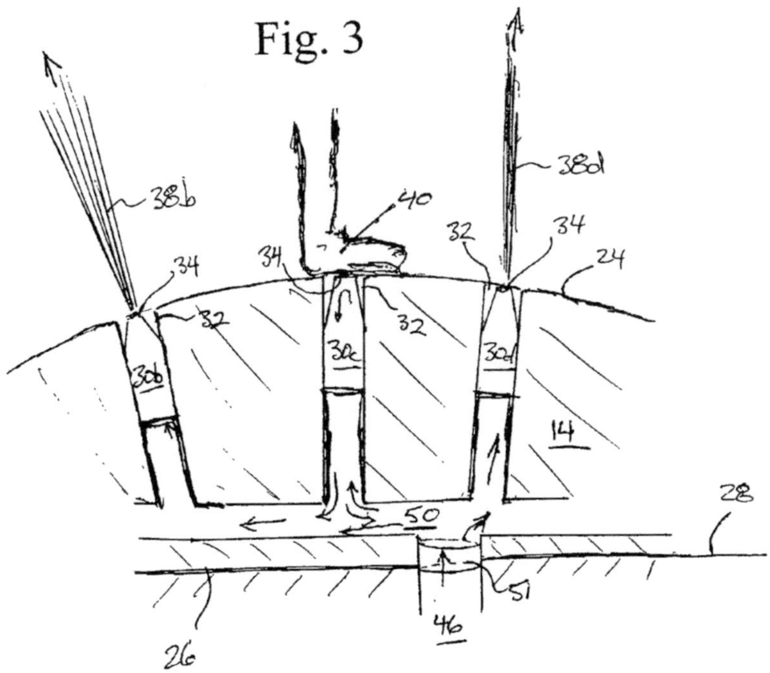

Fig. 7.2 Fig. 3 of US20030073505A1

Interaction is therefore achieved solely by closing an opening, causing water to spray from other openings at higher pressure. From this state of the art, one could infer the disadvantage that the device is limited in its play variations and that the fun may diminish relatively quickly.

A technical objective derived from the state of the art could therefore be to provide a water park amusement device characterized by a wide range of possible variations.

In the state of the art, EP 1898097 A2 ("Water pump and water play system with pump") can also be identified.

Figure 7.3 shows a section of Fig. 2 from EP 1898097 A2, in which valves 51 control the water supply to nozzles 50, with the valves 51 being operated via electrical lines 55. There is a central controller 57 for the electrical lines 55, so that each nozzle 50 can be controlled individually. This enables a wide range of variations in the control of the water fountains.

A new technical objective for the in-house technology can be formulated as increasing the attractiveness of the water features by combining the device with lighting elements, so that the water fountains are illuminated in color.

Conventional system (3kW)

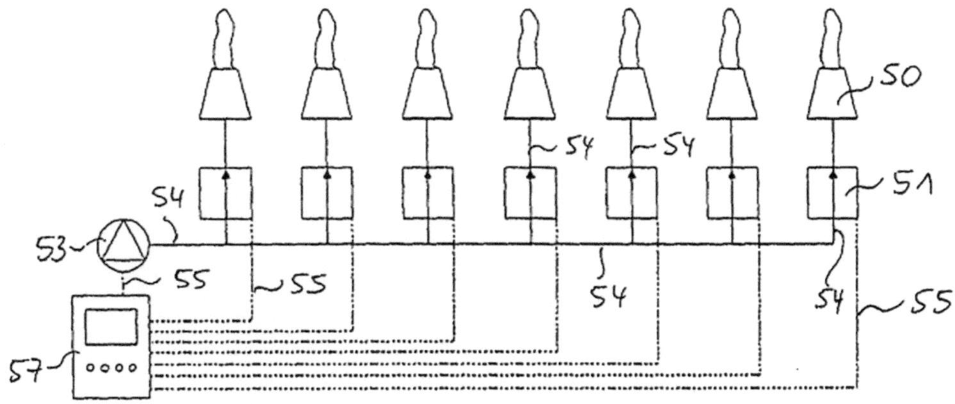

Fig. 7.3 Fig. 2 of EP1898097A2

However, lighting elements are also described in further prior art. US 4892250 (title: "Dynamic fountain displays and methods for creating the same") discloses in claim 3 a "*Patio fountain of claim 1 further including means for illuminating the water fountain display with colored light*".[2] Claim 3 therefore discloses means for illuminating the water fountains with colored light. In addition, US 4892250 shows valves 30 for each nozzle 34 of the arrangement.

Figure 7.4 shows Fig. 3 of US 4892250, in which individual valves 30 control water nozzles 34.

Figure 7.5 shows a computer 136 that controls the individual valves 140 of the water nozzles.

Claim 6 of US 4892250 reads: "*6. The patio fountain of claim 1 further comprised of a plurality of controllable valves and control means for controlling said controllable valves, each of said controllable valves being located between said pump means and a respective said nozzle, whereby the flow of water to each nozzle may be independently controlled.*"[3]

Claim 6 describes controllable valves, allowing the water flow to each nozzle to be regulated independently of the other nozzles, thereby enabling a high degree of variation in the control of the water nozzles.

The illumination of water fountains is also already described in the prior art. All features listed so far are therefore anticipated by the prior art, with the relevant patents

[2] DPMA, https://depatisnet.dpma.de/DepatisNet/depatisnet?action=pdf&docid=US000004892250A&xxxfull=1, accessed on 7.8.2024.

[3] DPMA, https://depatisnet.dpma.de/DepatisNet/depatisnet?action=pdf&docid=US000004892250A&xxxfull=1, accessed on 8.8.2024.

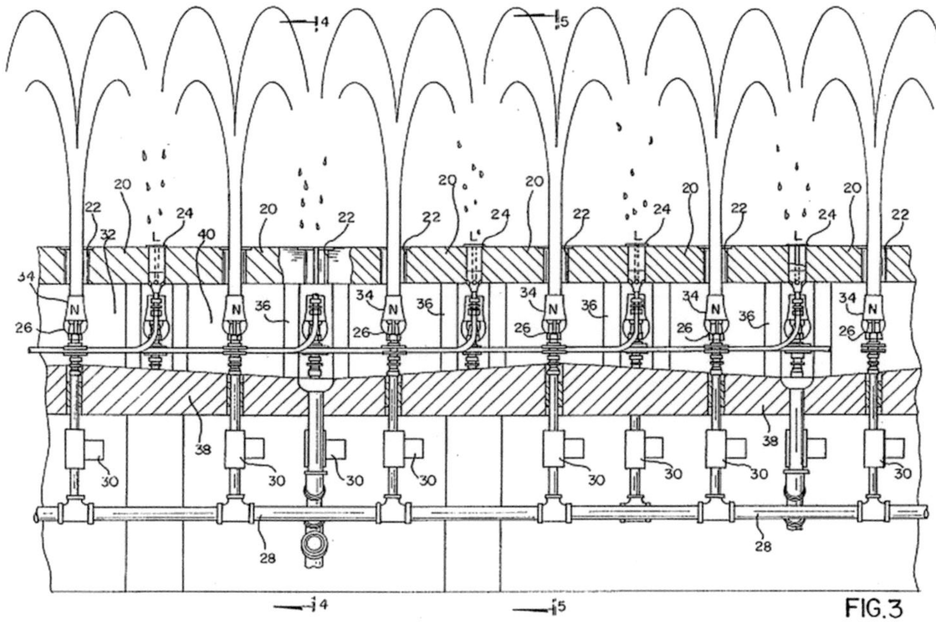

Fig. 7.4 Fig. 3 of US4892250

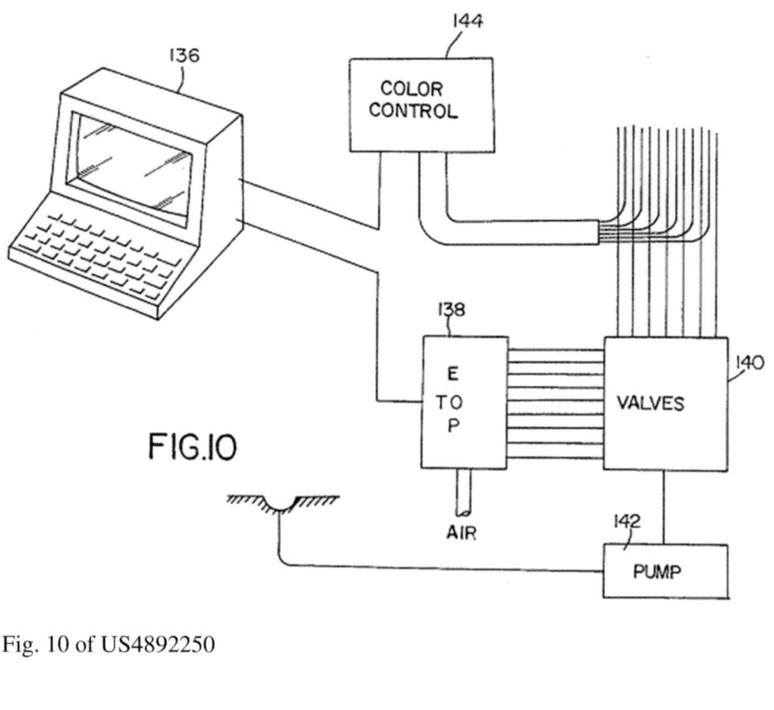

Fig. 7.5 Fig. 10 of US4892250

already disclosing the features of independent control of the water nozzles and illumination of the water fountains. The technology considered so far is therefore in the public domain, as US20030073505A1 and US 4892250 were filed more than 20 years ago, meaning the maximum patent term of 20 years has already expired. A person skilled in the art would also combine these two patent documents, so a patent that merely combines the technical teachings of these two documents would not be legally valid. The technology developed so far can therefore be used without infringing any third-party patent. However, it does not constitute a new and inventive technology, so patentability is not given.

Objective of the technical invention
After evaluating the currently available technology, further development can be pursued to create an independent product that provides the startup with a competitive advantage through improved features.

For example, a water park device could be provided that can be quickly assembled and disassembled and is also adaptable to local conditions. Such a device would be suitable for mobile use and could be adjusted to fit the available space.

This technical objective could, for example, be achieved by aiming for a modular design, with each module already containing all the necessary elements. However, EP 2588252 B1 (title: "*Play installation with fountains*") must be taken into account.

The main claim of EP 2588252 B1 reads:

> *"1. Arrangement of a play fountain, comprising: a floor (1), several nozzles (2), preferably a collection tank for a liquid, in particular water, or a device for connecting the arrangement to a liquid supply, and a control system (3) for controlling one or more of the nozzles (2) during operation of the arrangement, characterized in that the floor (1) comprises several mutually removable floor modules (20), each of which has at least one nozzle (2) and at least one pump (22)."*[4]

The claim describes floor modules that include a nozzle and a pump. The claim therefore already discloses a complete module that can be combined with others to create a play area with water fountains.

Figure 7.6 shows modules 20 of EP 2588252 B1, which comprise a nozzle and a pump 22. However, it can also be seen that modules 20 require a structure 30 to accommodate the individual modules 10. Therefore, to set up the play area, the frame 30 must first be assembled, which predetermines the dimensions of the play area. In particular, the play area cannot be easily adapted to specific spatial conditions such as corners or

[4] DPMA, https://depatisnet.dpma.de/DepatisNet/depatisnet?action=pdf&docid=EP000002588252B1&xxxfull=1, accessed on 8.8.2024.

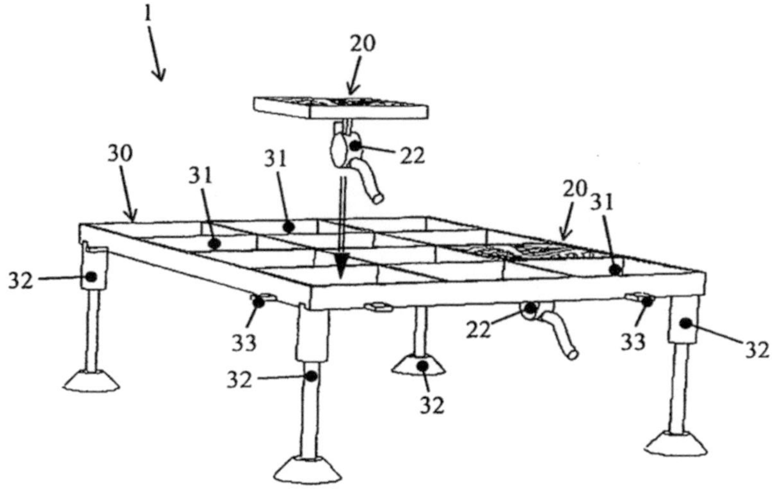

Fig. 7.6 Fig. 2 of EP2588252B1

sloped surfaces. Rather, at least enough space must be available for the rectangle of the frame 30.

A specific technical objective, therefore, would be to provide a play area with water fountains that can be adapted to a given, potentially challenging, spatial situation.

Creation of the Invention

Figure 7.7 shows an invention that results from the examination of the prior art. According to the invention, a cube is provided that can be placed on the ground without a frame and is equipped with connectors 3 and 4. The cube includes a nozzle and a pump, so that a water fountain 2 can be ejected from the cube in a controlled manner. Connector 3 allows the cube to be connected to an electrical supply and control line, while connector 4 is used for water supply. The cube has connectors 3 and 4 on all four sides, allowing it to be connected to adjacent cubes. For end cubes, unused water connections can be sealed with plugs.

The inventive module can therefore be used without a frame structure and, depending on the available space, the modules can be arranged in any desired configuration, with the smallest predefined unit of space corresponding to one inventive module.

Specialized embodiments for specific applications

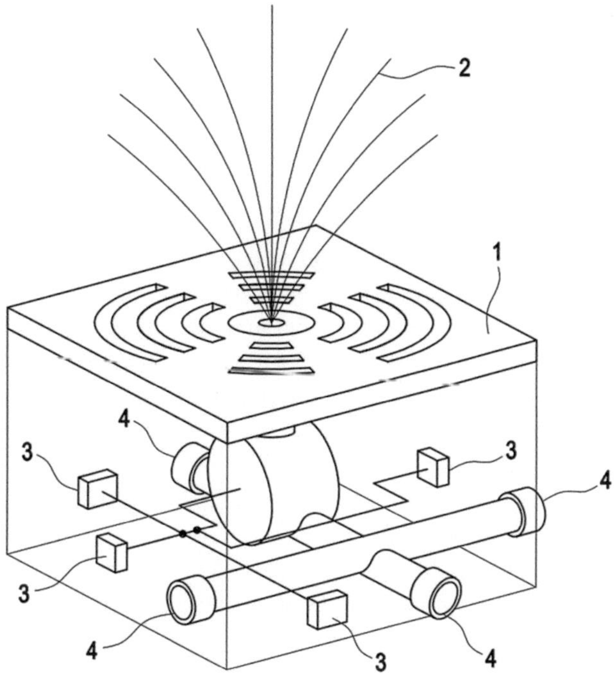

Fig. 7.7 Module of a play area

Once the fundamental technical concept has been established, special embodiments can be considered. For example, the inventive cube could be equipped with lighting elements that illuminate the cube itself and/or the water fountain.

7.3 Freedom-to-Operate

In a freedom-to-operate assessment, it must be determined whether one's own product infringes the scope of protection of third-party patents. Assessing patent applications is difficult, as it is usually not possible to make a clear statement about the scope of protection after grant. It is also possible that the patent application may be rejected in its entirety.

A freedom-to-operate opinion is only as good as the preceding search. A newly discovered document can overturn the result of a freedom-to-operate opinion that a product does not infringe third-party patents.

However, there is an exception. If all features of a product can be found in a document that is already in the public domain, the product can be manufactured without a newly discovered document being able to restrict this right. This right arises from a document that is already in the public domain and can no longer be diminished. A public domain document is, in particular, a patent whose maximum term of protection has expired and which therefore no longer confers an exclusive right.

7.4 Patentability of One's Own Product

The question of the patentability of one's own product can be easily answered if all features of the product are contained in one of the prior art documents. In this case, the product is not novel and patent grant is excluded.

If the features of one's own product can be found in a combination of several documents, it must still be determined whether a person skilled in the art would combine the relevant documents. If so, the product lacks inventive step. However, if the features of one's own technology cannot be derived from any combination of the prior art documents, the product is at least considered patentable over the prior art that was searched.

The problem with assessing patentability is that a positive assumption of patentability can never be made. There is always the possibility that a document may emerge that destroys patentability. Conversely, if a lack of patentability has been established, this result cannot be overturned by newly discovered documents. Therefore, only a lack of patentability can be positively established.

8 Examples

Contents

Using concrete examples, the individual steps involved in developing a technology for a startup are explained. In particular, the application of the state of the art is presented in order to assess and further develop one's own technology.

8.1 Weighing Unit on a Shopping Cart

The checkout process in supermarkets is increasingly being automated to enable convenient and fast payment. In particular, customers are being offered the option to self-scan their items. One idea is to arrange a weighing unit on the shopping cart in such a way that the items in the cart can be weighed. This allows the supermarket to check whether all and the correct items have been scanned during self-checkout, as a comparison with the weight of the items is made possible.

A technical concept, therefore, is a shopping cart that provides a weighing unit for the items in the basket and on the lower tray. In particular, a single weighing unit should be usable for both the items in the basket and on the lower tray, so that the technical complexity of the device can be kept low.

A search of the prior art yielded the following documents:

© The Author(s), under exclusive license to Springer-Verlag GmbH, DE, part of Springer Nature 2025

T. H. Meitinger, *Startup Inventor's Handbook*,
https://doi.org/10.1007/978-3-662-72134-6_8

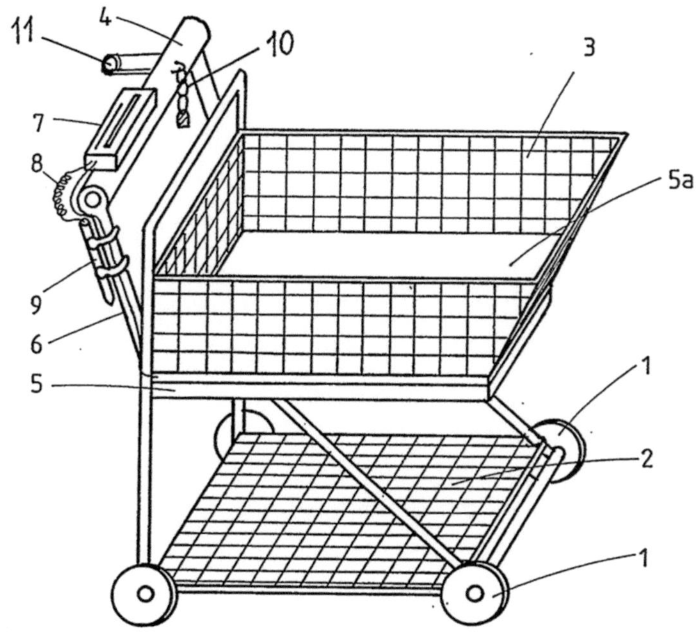

FIG.1

Fig. 8.1 Fig. 1 of DE19643122A1

D1: DE 19643122 A1, filing date: October 18, 1996
D2: US 2011/0036907 A1, filing date: August 10, 2010
D3: DE 10 2018 132 059 B4, filing date: December 13, 2018

Patent application D1 describes a shopping cart suitable for use in a self-service store.

Figure 8.1 shows Fig. 1 of D1 with a basket 3, under the base 5a of which a scale 5 is arranged. The scale 5 can be used to weigh the items in the basket 3. The customer can also stack items, such as beverage crates, on the lower tray 2. With the embodiment shown in Fig. 8.1, these items cannot be weighed. However, the patent application contains the following passage: *"Preferably, however, the lower tray also forms the support surface of a further electronic scale, which is connected to a display device."*[1]

[1] DPMA, https://depatisnet.dpma.de/DepatisNet/depatisnet?action=pdf&docid=DE000019643122A1&xxxfull=1, column 4, lines 5 to 7 of the published application.

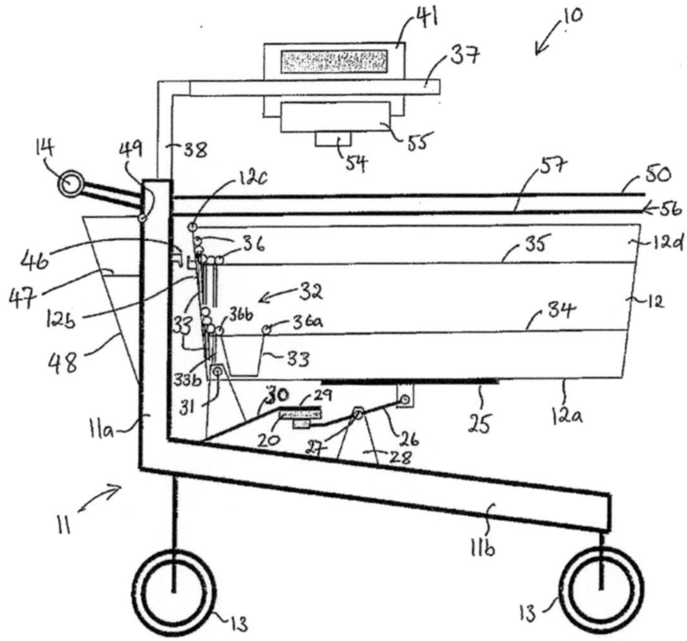

Figure 1b

Fig. 8.2 Fig. 1b of US20110036907A1

Document D1 therefore describes the weighing of items in the upper basket and on the lower tray. However, the document does not disclose how this can be achieved with only a single scale.

Document D2 (US 20110036907 A1) describes a shopping cart with several separate baskets.

Figure 8.2 shows a shopping cart with several baskets 33, some of which are stacked on top of each other and whose contents can be weighed by only one scale 20.[2] However, document D2 does not show the classic version of a shopping cart with an upper basket and a lower tray. Instead, the shopping cart in D2 contains individual foldable baskets 33.

[2] DPMA, https://depatisnet.dpma.de/DepatisNet/depatisnet?action=pdf&docid=US020110036907A1&xxxfull=1, accessed on 16.8.2024.

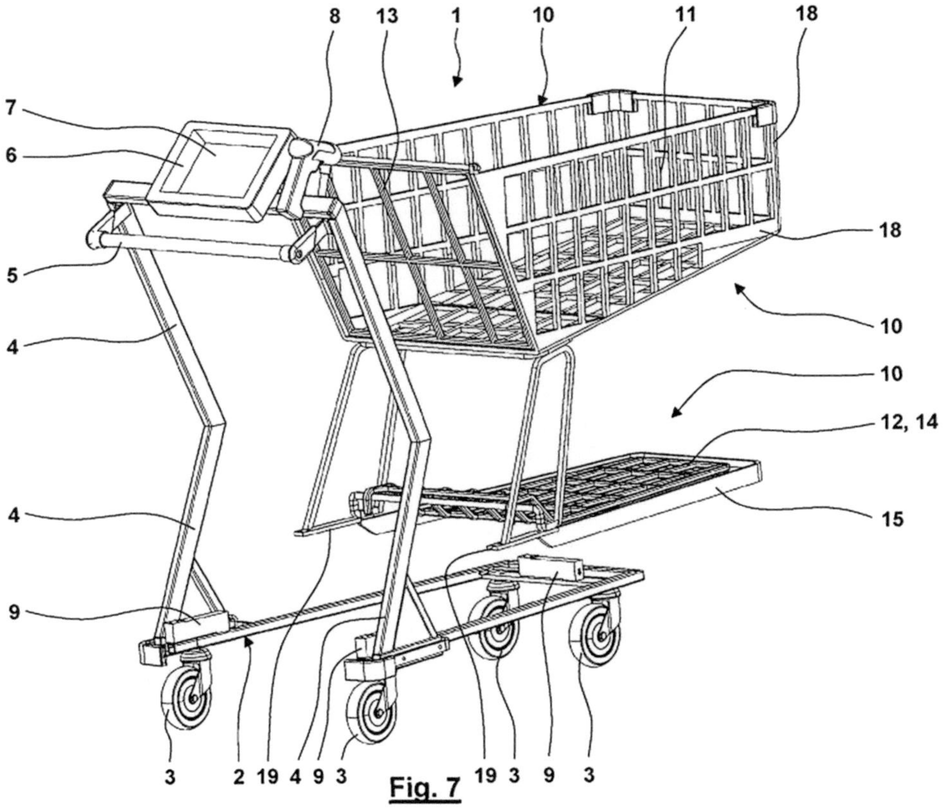

Fig. 8.3 Fig. 7 of DE102018132059B4

Patent D3 (DE 10 2018 132 059 B4) discloses an upper basket 18 and a lower tray 15.[3]

Figure 8.3 shows Fig. 7 of patent D3 with a wire basket 18 and a lower tray 15, which are rigidly connected to each other and rest on one or more load cells 9. Due to the rigid connection of the upper basket 18 with the lower tray 15, it is possible to weigh both the items in the basket 18 and the items on the tray 15 with only one scale.

Patent D3 (DE 10 2018 132 059 B4) is still in force. The use of the technical teaching of D3 may therefore be prohibited. A workaround would be to separate the upper basket from the lower tray and to mount the basket and the tray directly on a load cell 9. In this

[3] DPMA, https://depatisnet.dpma.de/DepatisNet/depatisnet?action=pdf&docid=DE102018132059B4&xxxfull=1, accessed on 13.2.2025.

way, patent D3 would be circumvented and both the items in the basket 18 and on the lower tray 15 could be weighed with only one scale 9.

8.2 Balcony for Retrofitting

A kit for retrofitting a balcony to an existing house is to be made available on the market. The floor of the balcony could additionally be used as the roof of a carport or garage.

A provisional application was drafted, and a request for examination was submitted simultaneously with the filing of the patent application, so that the patent office promptly determined the relevant prior art.[4] Currently, the fee for filing a request for examination is €350, which is a very reasonable price for a prior art search.[5]

The main claim of the provisional application read as follows:

> *"1. Kit for constructing a balcony comprising:*
> *one, two, three, four, five, or any number of supports (2),*
> *a floor (3), wherein the support (2) holds the floor (3), and*
> *a railing (1) to protect against a person falling from the balcony,*
> *wherein the railing (1) is connected to one or two supports (2), characterized in that the balcony is suitable for retrofitting to an existing building."*[6]

It was also described that the kit can additionally or alternatively be used as a carport:

> *"The above also applies, in another aspect, not only to balconies but analogously and correspondingly to carports. Thus, a further invention relates to a kit for constructing a carport comprising: one, two, three, four, five, or any number of supports, a floor, wherein the support holds the floor, and a railing to protect against a person falling from the carport, wherein the railing is connected to one or two supports, and the carport is suitable for retrofitting to an existing building."*[7]

Figure 8.4 shows an inventive kit with supports 2, a floor 3, and a railing 1.

The search conducted by the German Patent Office identified several documents describing the retrofitting of balconies. The basic idea was therefore well known and would hardly be able to secure any significant market advantage. At the very least, it could be assumed that the planned business idea constitutes prior art in the public

[4] Section 43, paragraph 1, sentence 1 of the Patent Act.

[5] Fee item 311.400 according to the annex to Section 2, paragraph 1 of the Patent Costs Act (Schedule of Fees).

[6] DPMA, https://depatisnet.dpma.de/DepatisNet/depatisnet?action=pdf&docid=DE102016118690A1&xxxfull=1, accessed on 15.8.2024.

[7] DPMA, https://depatisnet.dpma.de/DepatisNet/depatisnet?action=pdf&docid=DE102016118690A1&xxxfull=1, accessed on 15.8.2024.

Fig. 8.4 Fig. 1 of DE102016118690

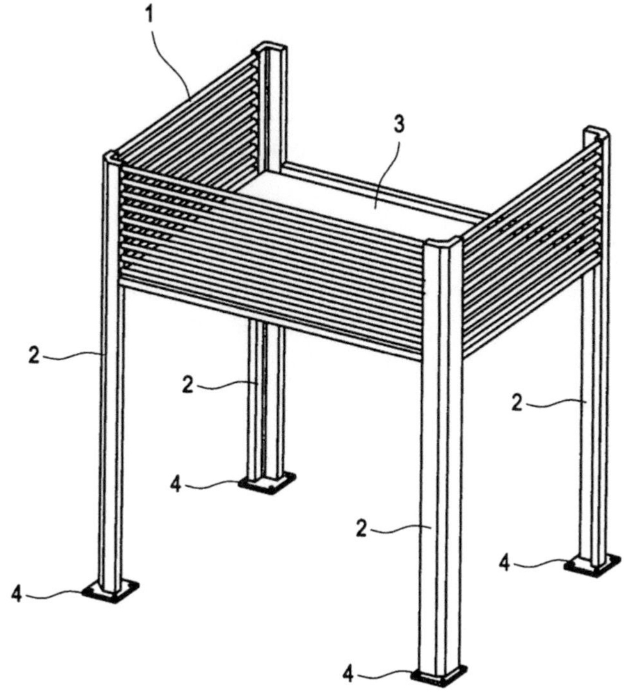

Fig. 1

domain and can therefore be used without infringing third-party intellectual property rights.

Specifically, the search yielded the following documents:

D1: DE 20 2014 000 276 U1, filing date January 16, 2014
D2: DE 297 16 050 U1, filing date September 8, 1997
D3: DE 298 06 721 U1, filing date April 6, 1998
D4: DE 94 09 626, filing date June 15, 1994
D5: GB 2356 409 A, filing date October 29, 1999
D6: DE 3210215 A1, filing date March 19, 1982

Document D1 describes in its main claim a balcony system for retrofitting. The main claim reads:

"1. Balcony system (10), in particular for retrofitting projecting balconies to existing buildings, with four posts (12, 14, 16, 18) supported on a base, in particular a ground foundation, and positioned in front of a building, and one or more balcony slabs (20) arranged one above the other by floor, wherein each balcony slab (20) has a substructure (74) and a

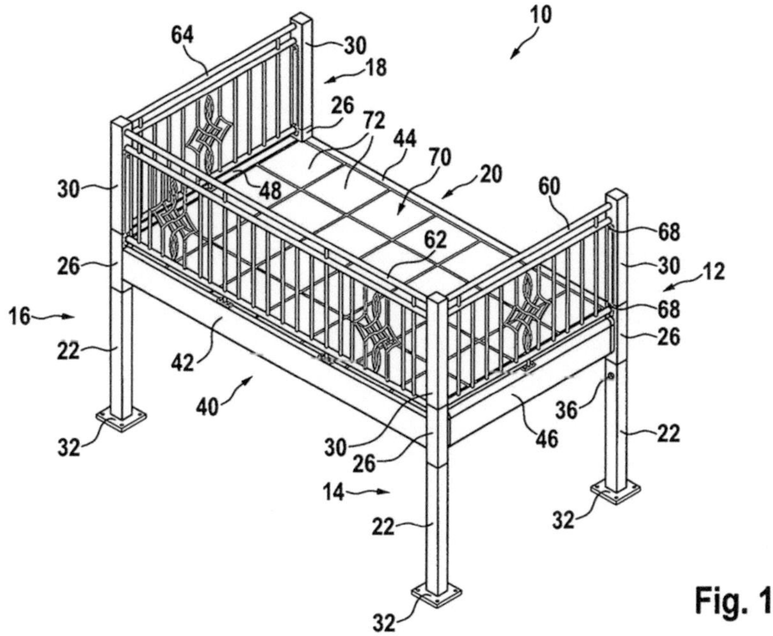

Fig. 8.5 Fig. 1 of DE202014000276U1

*deck (70) on top of it, and for each balcony slab, a supporting frame (40) surrounding it is held on the posts,**characterized in that**the supporting frame (40) comprises a horizontally extending front frame part (42) and a parallel, building-facing rear frame part (44), each having at both longitudinal ends a preferably fixedly connected, in particular welded, vertically extending socket (26) with an upper open end face (28) for a plug connection with one of the posts (12; 14; 16; 18), as well as two horizontally extending side frame parts (46, 48), which are hooked in at the respective longitudinal ends between the front frame part (42) and the rear frame part (44) and are detachably secured thereto, and the substructure (74) between the front frame part (42) and the rear frame part (44) comprises at least two horizontally extending crossbeams (76, 78, 80, 82, 84) parallel to each other and at right angles to the rear frame part, whose longitudinal ends are frictionally connected to the front frame part (42) and the rear frame part (44), as well as at least one longitudinal beam (86) extending transversely to the crossbeams and laid horizontally on them, wherein the deck (70) rests on the supporting frame (40) and the at least one longitudinal beam (86)."*[8]

Figure 8.5 shows a mountable balcony system that is very similar to the balcony in Fig. 8.4.

[8] DPMA, https://depatisnet.dpma.de/DepatisNet/depatisnet?action=pdf&docid=DE202014000276U1&xxxfull=1, accessed on 15.8.2024.

D3 also discloses a very similar technical teaching of an attachable balcony system. The main claim of utility model D3 reads as follows:

*"1. Balcony construction for retrofitting to existing exterior building walls using metal hollow profiles as supports and open, rigidly joined frames mitered together, which are reinforced with cross braces, with floor panels supported on the frame and the cross braces,**characterized in that,**on the frame (4), which consists of an upwardly open G-shaped metal profile, a floor panel (6) provided with a bevel (16) is supported, with a gap (20) running around the entire floor panel (6) arranged between the bevel (16) and the frame (4), and that two connecting sleeves (2) are attached to each narrow side of the frame (4), from which a lower connecting element (17) and an upper connecting element (18) respectively protrude, so that the metal hollow profiles of the lower and upper connecting elements (17; 18) correspond in a slidable manner with the hollow profiles of the supports (1)."*[9]

Figure 8.6 shows an attachable balcony system according to D3.

The technical teachings of the prior art disclose exclusively supports of the respective balcony systems that have a square cross-section. In the present application, a support is described that has a special configuration, as it is characterized by three sections, each offset by 45°.[10]

Figure 8.7 shows in Fig. 3 an inventive support 2, which comprises the sections A, B, and C, arranged at angles. The advantage of this design is that the support is more stable, results in a more attractive design, and, despite high mechanical stability, has a low weight.

A new main claim to distinguish from the prior art could therefore be formulated in particular by incorporating the features of dependent claim 2.

Dependent claim 2 reads as follows:

"2. Kit according to claim 1, characterized in that the support (2) consists of segments (A, B, C) and/or the segments (A, B, C) are arranged at a 45° angle."[11]

8.3 Bicycle with High Riding Comfort

The fundamental idea is to provide a bicycle that allows the rider to sit comfortably even on poor road surfaces or off-road. It was recognized that, for this purpose, the seat should not be mounted vertically on the bicycle, but horizontally, so that the seat is able to swing or oscillate, thereby absorbing shocks and preventing them from being transmitted directly to the rider.

[9] DPMA, https://depatisnet.dpma.de/DepatisNet/depatisnet?action=pdf&docid=DE000029806721U1&xxxfull=1, accessed on 15.8.2024.

[10] DPMA, https://depatisnet.dpma.de/DepatisNet/depatisnet?action=pdf&docid=DE102016118690A1&xxxfull=1, accessed on 28.9.2025.

[11] DPMA, https://depatisnet.dpma.de/DepatisNet/depatisnet?action=pdf&docid=DE102016118690A1&xxxfull=1, accessed on 15.8.2024.

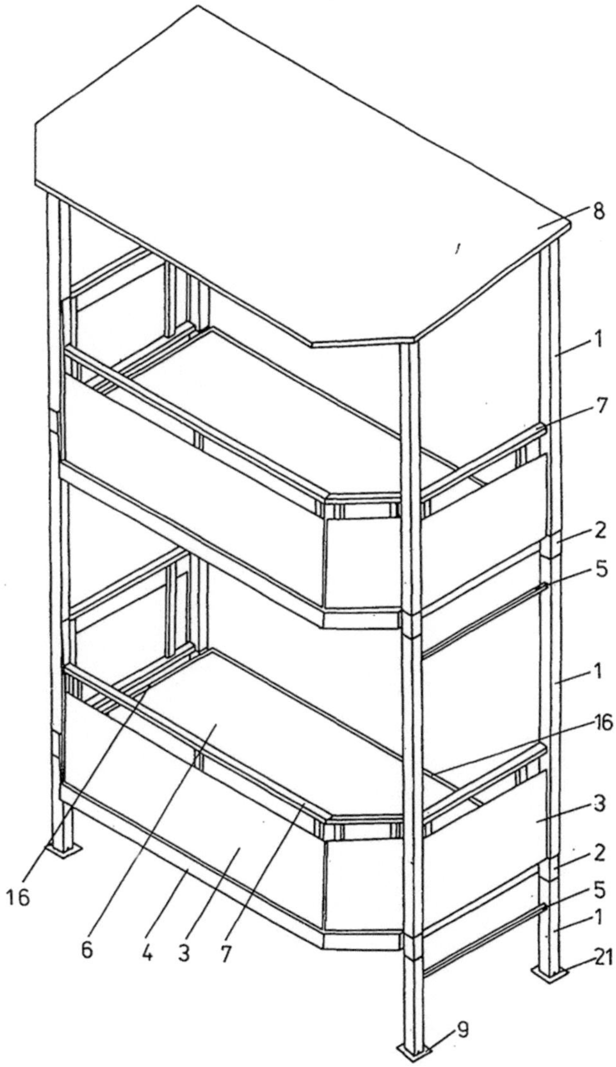

Fig. 8.6 Fig. 2 of DE29806721U1

Figure 8.8 shows, at the top, the conventional mounting of a bicycle seat on a bicycle. The illustration below shows the horizontal mounting of the bicycle seat.

A search of the prior art yielded the following documents:

D1: US 5240268, filing date August 31, 1993
D2: US 2244709, filing date November 15, 1938
D3: FR 820740, filing date July 24, 1936
D4: FR 2807995, filing date April 19, 2000

For all documents of the prior art, the maximum patent term of 20 years has already expired, so the technical teachings of these patent documents are in the public domain and may be used by anyone without restriction.

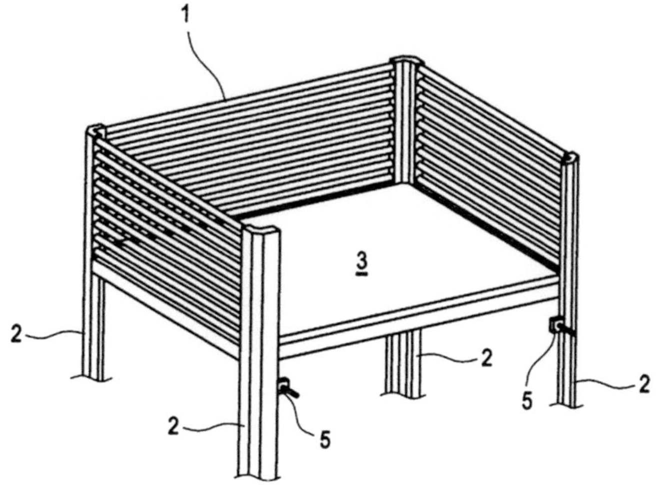

Fig. 2

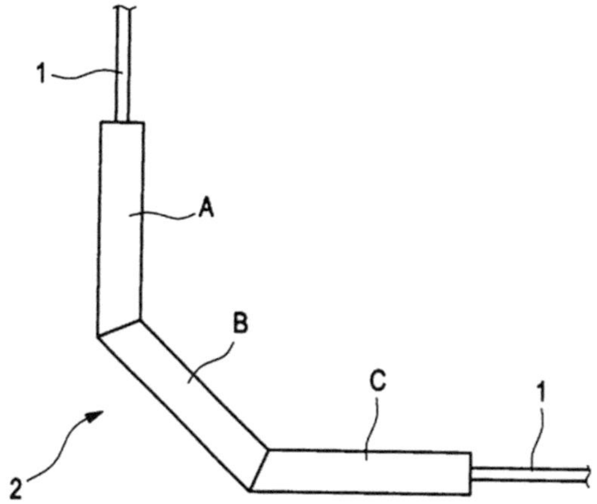

Fig. 3

Fig. 8.7 Figs. 2 and 3 of DE102016118690

Document D1 (US 5240268) describes a bicycle featuring a horizontal mounting of the bicycle seat. Figure 8.9 shows the bicycle from D1 with the bicycle saddle 26, which is attached to a tube 18 that is formed almost horizontally.[12]

[12] DPMA, https://depatisnet.dpma.de/DepatisNet/depatisnet?action=pdf&docid=US000005240268A&xxxfull=1, abgerufen am 28.9.2025.

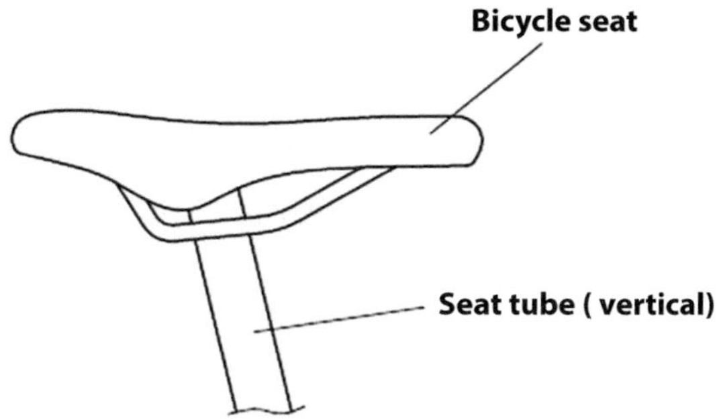

Fig. 8.8 Comfortable bicycle seat

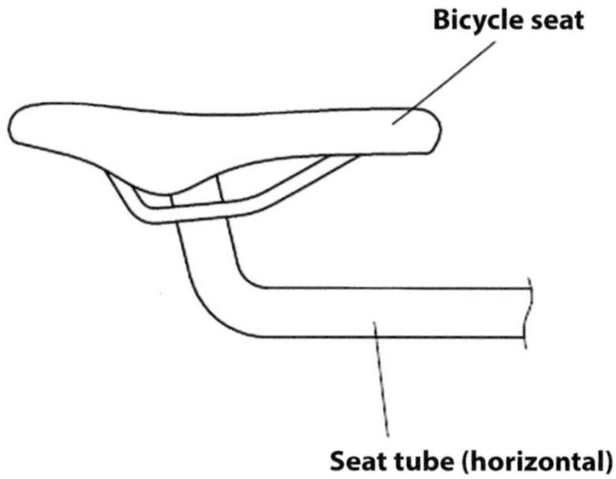

Figure 8.10 shows an alternative embodiment according to document D1 (US 5240268), in which the bicycle seat is also mounted in an almost horizontal orientation on the bicycle.[13]

Patent D2 (US 2244709) describes another bicycle with a horizontally suspended bicycle saddle. Figure 8.11 shows the bicycle saddle according to the invention.[14]

The disclosure of document D3 (FR 820740) corresponds to the technical teachings of D1 and D2. Figure 8.12 shows the bicycle from D3.[15] In Fig. 8.12, it is evident that a

---

[13] DPMA, https://depatisnet.dpma.de/DepatisNet/depatisnet?action=pdf&docid=US000005240268A&xxxfull=1, accessed on 19.8.2024.

[14] DPMA, https://depatisnet.dpma.de/DepatisNet/depatisnet?action=pdf&docid=US000002244709A&xxxfull=1, accessed on 19.8.2024.

[15] DPMA, https://depatisnet.dpma.de/DepatisNet/depatisnet?action=pdf&docid=FR000000820740A&xxxfull=1, accessed on 19.8.2024.

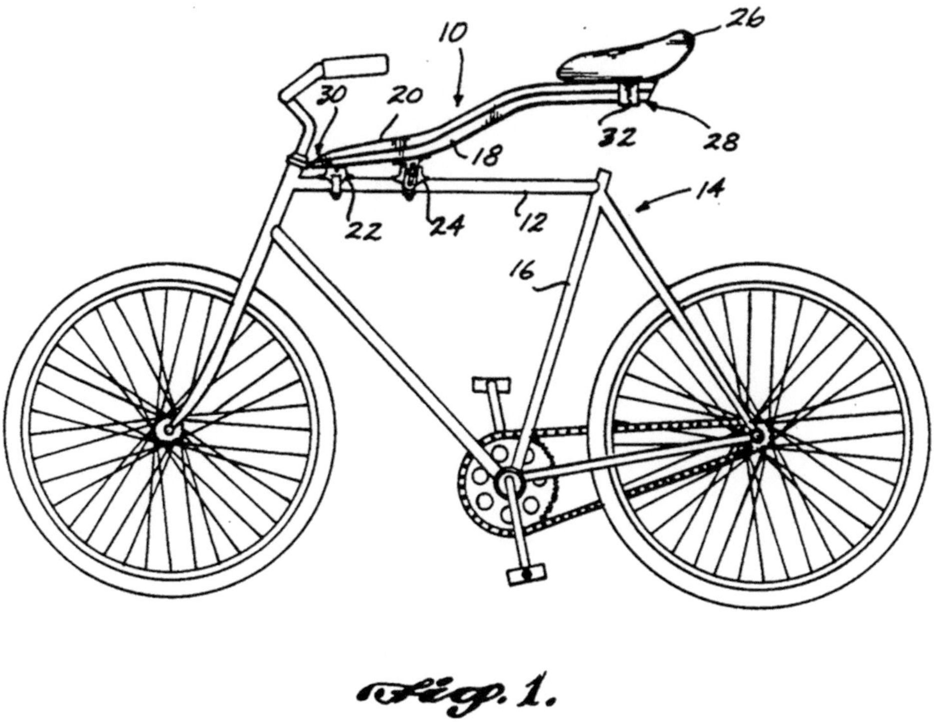

Fig. 8.9 Fig. 1 of US5240268

freely suspended, horizontal mounting of the bicycle saddle can lead to oscillations with large amplitude, which is detrimental to riding comfort.

Figure 8.13 shows Fig. 1 of patent D4 (FR 2807995) with an improved version of an almost horizontal mounting of the bicycle seat. Figure 8.13 shows an arrangement of a bicycle seat, where the frame 1 acts as a damping element to prevent excessive oscillation of the bicycle seat 8.[16]

Thus, the prior art has already anticipated the horizontal arrangement of the bicycle saddle on the bicycle. It has also been recognized that damping would be beneficial to prevent excessive oscillation of the bicycle seat.

As a proprietary technology, damping of the oscillation of a horizontally oriented bicycle seat could be achieved by means of an eddy current brake, which could also advantageously generate electrical energy.

[16] https://depatisnet.dpma.de/DepatisNet/depatisnet?action=pdf&docid=FR000002807995A1&xxxfull=1, accessed on 13.2.2025.

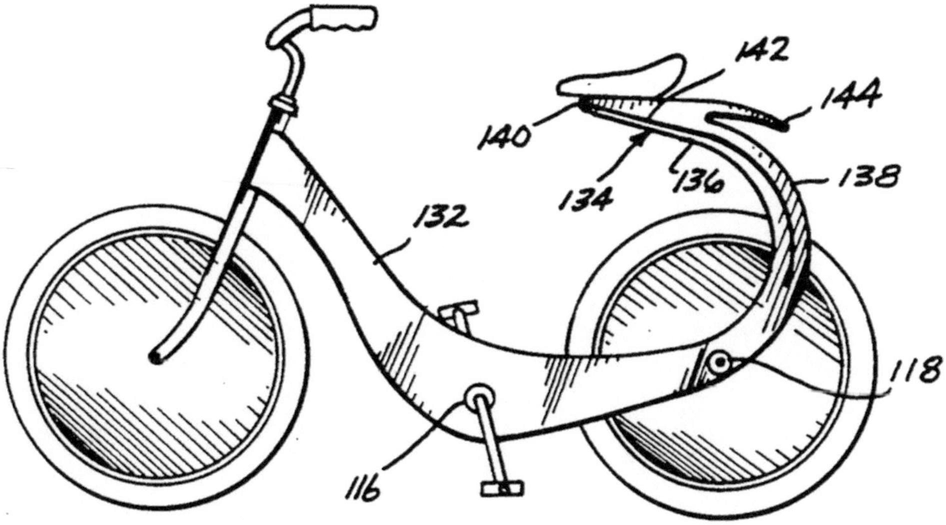

Fig. 10.

Fig. 8.10 Fig. 10 of US5240268

Fig. 1.

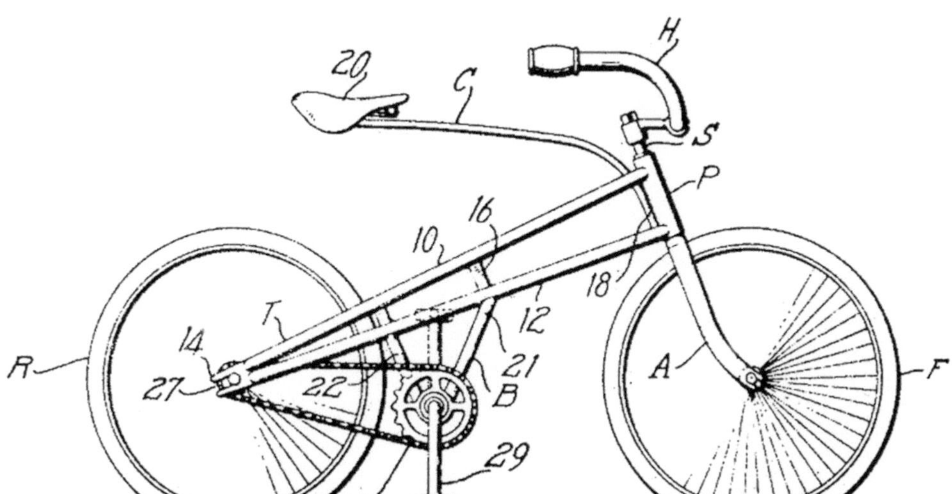

Fig. 8.11 Fig. 1 of US2244709

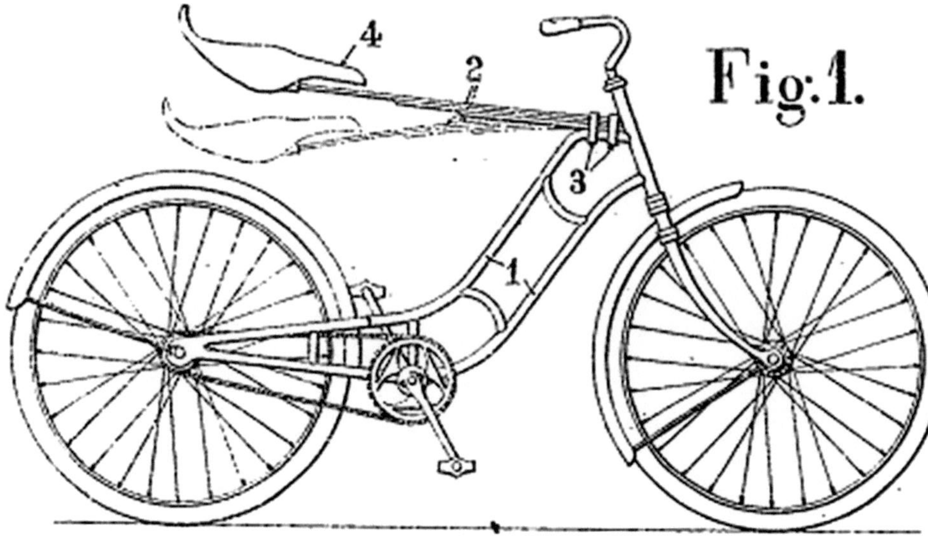

Fig. 8.12 Fig. 1 of FR820740

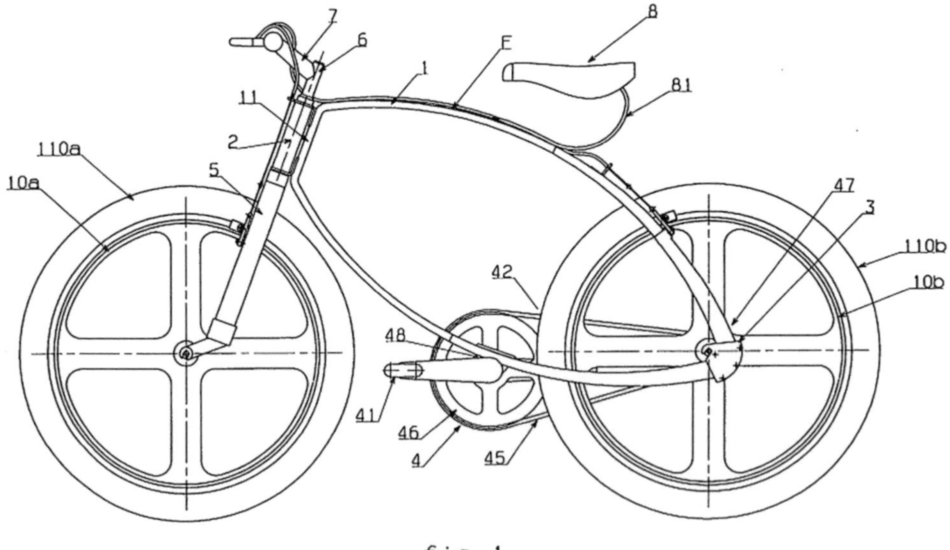

Fig. 8.13 Fig. 1 of FR2807995

8.4 Fasteners for Building Structures

A fastener is to be provided for the assembly of building components, such as wall elements, in such a way that unintentional loosening of the installed fastener is prevented. A proprietary technology describing such a fastener is disclosed in EP 3995704 B1.

Figure 8.14 shows the proprietary fastener with a thread 6 and a conical spring 9, allowing the fastener to be screwed in and held under pressure by the spring 9.[17]

Figure 8.15 shows the fastener connecting two blocks.

Fig. 8.14 Fig. 1 of EP3995704B1 (vertical)

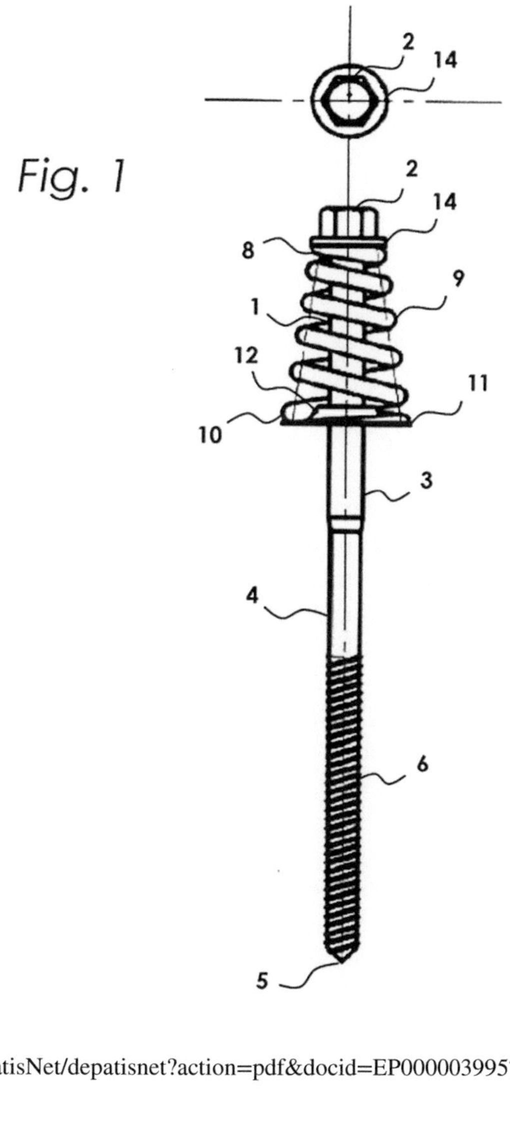

[17] DPMA, https://depatisnet.dpma.de/DepatisNet/depatisnet?action=pdf&docid=EP000003995704B1&xxxfull=1, accessed on 19.8.2024.

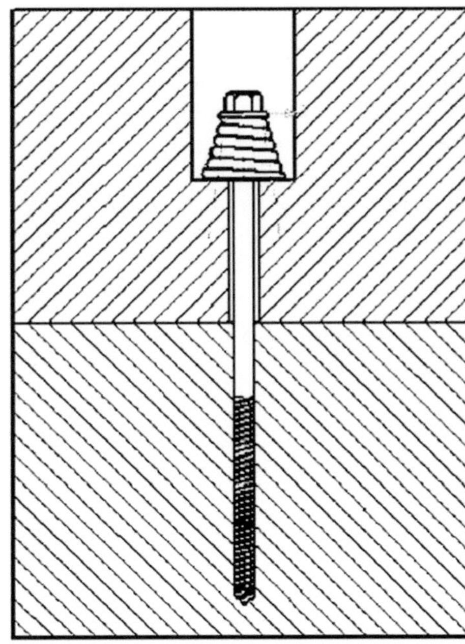

Fig. 8.15 Fig. 3 of EP3995704

As prior art, the patent office identified US2019/0055980. Figure 8.16 shows the cited screw from the prior art, which also features a spring for tensioning.[18]

A distinction from the prior art can be made in that the spring of the proprietary invention is left-handed, whereas in the prior art it is right-handed (see comparison of Fig. 8.16 as the cited prior art and 8.17 as the inventive fastener). The advantage is that when the screw is driven in, the spring is wound helically and thus does not lose its tension. In contrast, with the right-handed spring of Fig. 8.16, driving in the fastener causes the individual coils of the spring to stack on top of each other, resulting in a significant loss of tension.

8.5 Interlocking Furniture

Standard furniture often cannot make optimal use of the available space. Therefore, a technical challenge is to provide furniture components that can be assembled in such a way that a given space is utilized as efficiently as possible. This is particularly chal-

[18] DPMA, https://depatisnet.dpma.de/DepatisNet/depatisnet?action=pdf&docid=US020190055980A1&xxxfull=1, accessed on 28.9.2025.

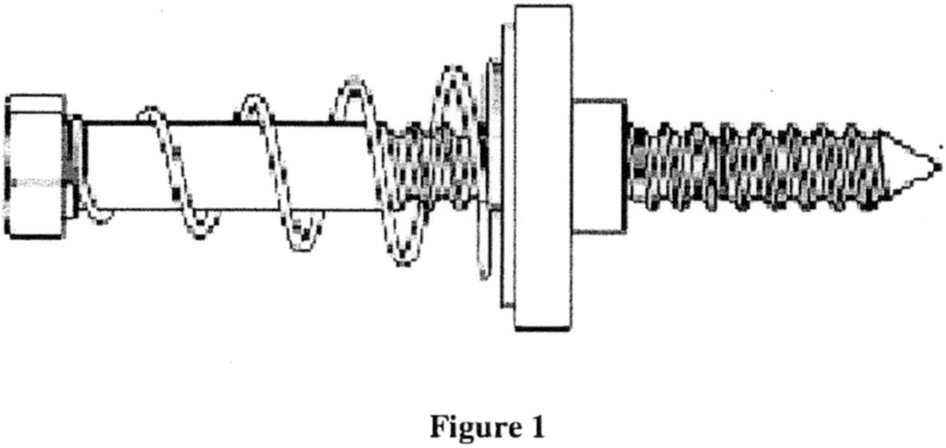

Fig. 8.16 Fig. 1 of US20190055980A1

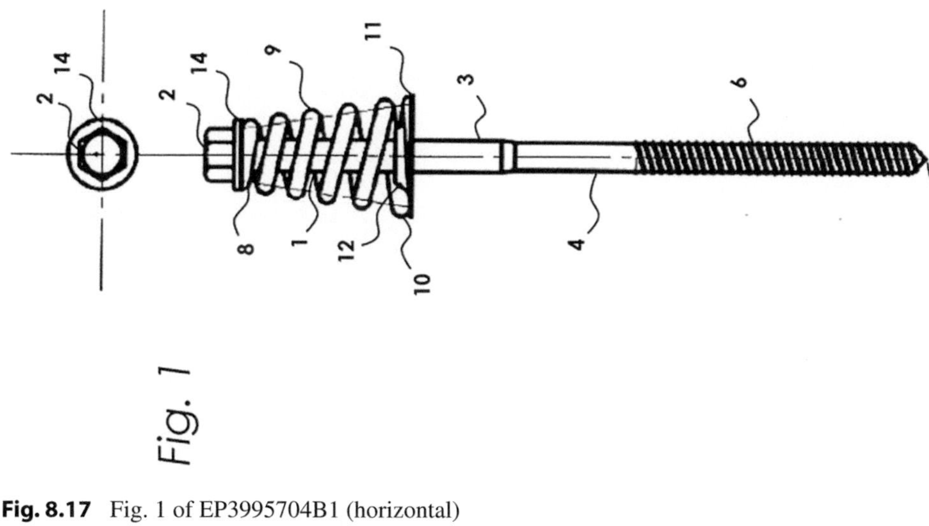

Fig. 8.17 Fig. 1 of EP3995704B1 (horizontal)

lenging in older houses or in the attic of a house, as these involve angled areas that are intended to be used as storage space with the furniture pieces.

As a starting point, a technical solution is chosen in which individual panels of different sizes are available, so that large panels allow for rapid assembly and small panels enable optimal use of angled areas. The individual panels are stably connected to each other by fasteners. The essential core of the technical solution is the fasteners.

Document EP 4302648 A1 describes a method for connecting furniture panels.[19]

[19] DPMA, https://depatisnet.dpma.de/DepatisNet/depatisnet?action=pdf&docid=EP000004302648A1&xxxfull=1, accessed on 31.8.2024.

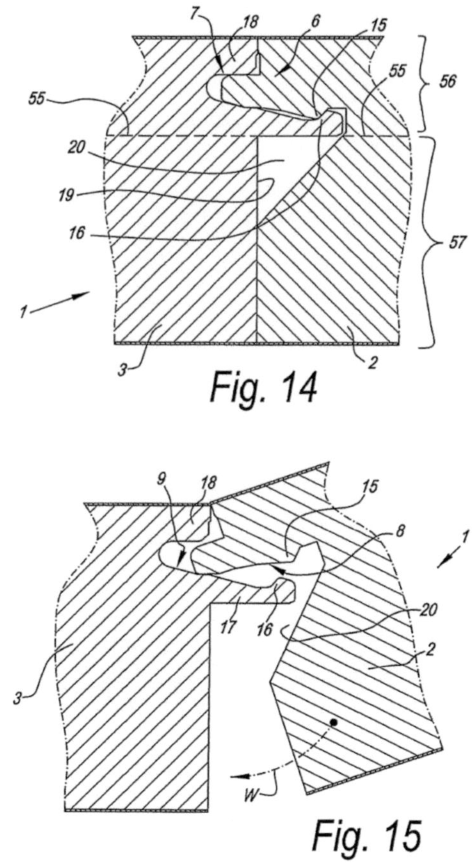

Fig. 8.18 Fig. 15 of EP4302648A1

Figure 8.18 shows connection options that require interlocking or undercut engagement. These are complex shapes whose production is time-consuming and expensive. It is also questionable whether the thin webs and small lugs can withstand high mechanical loads. Due to the complex structure of the fasteners, it can be assumed that not many fasteners will be used per panel, which raises doubts about the stability of the panel connections.

Document US 11821447 B2 also discloses fasteners for panels as prior art.[20]

Figure 8.19 shows a fastener from US 11821447 B2 with a two-part dovetail B or 2013. The fastener appears to have a complex structure.

[20] DPMA, https://depatisnet.dpma.de/DepatisNet/depatisnet?action=pdf&docid=US000011821447B2&xxxfull=1, accessed on 31.8.2024.

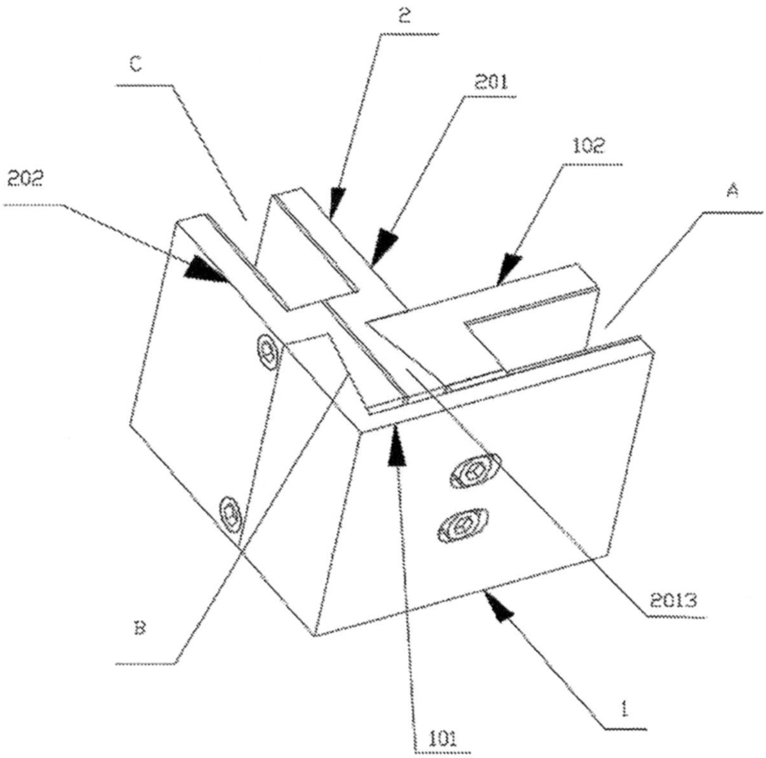

FIG.1

Fig. 8.19 Fig. 1 of US11821447B2

Figure 8.20 shows the fastener for joining two panels. The result of the connection does not produce flat surfaces. In addition, there are gaps between the panels through which items stored in the piece of furniture may fall.

The document EP 2527662 A1 from the prior art discloses fasteners that are joined together using clips.[21]

Figure 8.21 shows the fasteners of EP 2527662 A1. It can be seen that the connecting elements do not extend far into the respective panels, so the connection is unstable with respect to buckling.

Figure 8.22 shows the connecting elements with webs 130 and lugs 131 and 132, which are very thin and therefore fragile, so that a mechanically stable connection of two panels is not achieved.

---

[21] DPMA, https://depatisnet.dpma/DepatisNet/depatisnet?action=pdf&docid=EP000002527662A1&xxxfull=1, accessed on 31.8.2024.

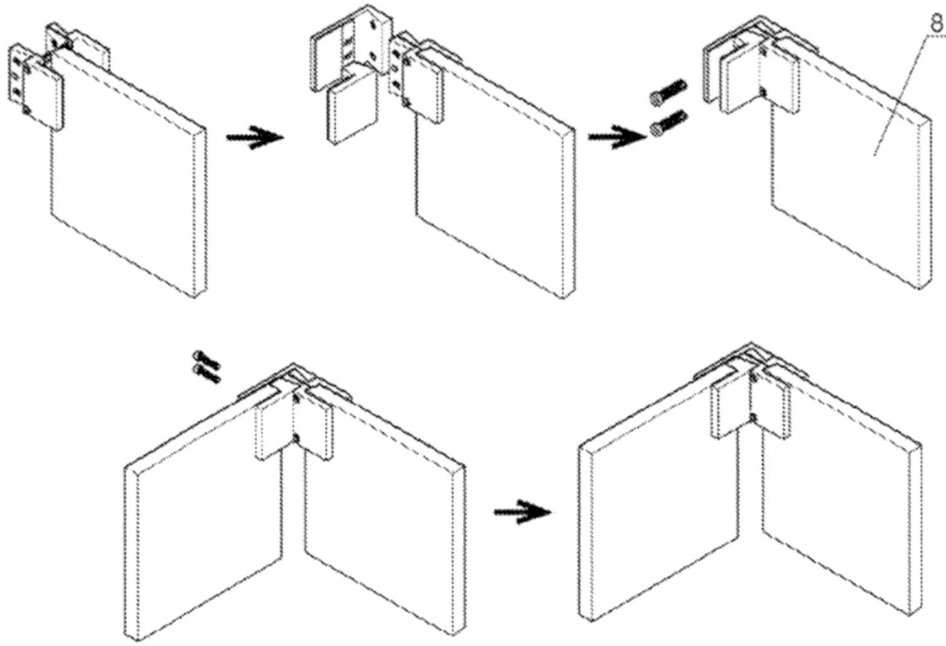

FIG. 6

Fig. 8.20 Fig. 6 of US11821447B2

Fig. 8.21 Fig. 1 of EP2527662A1

Fig. 1

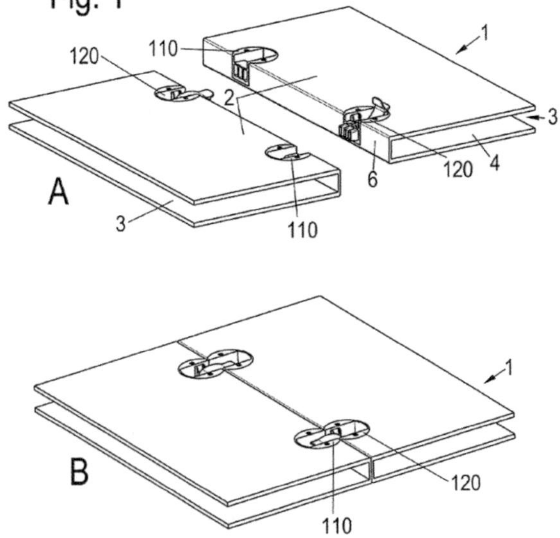

Fig. 8.22 Fig. 4 of EP2527662A1

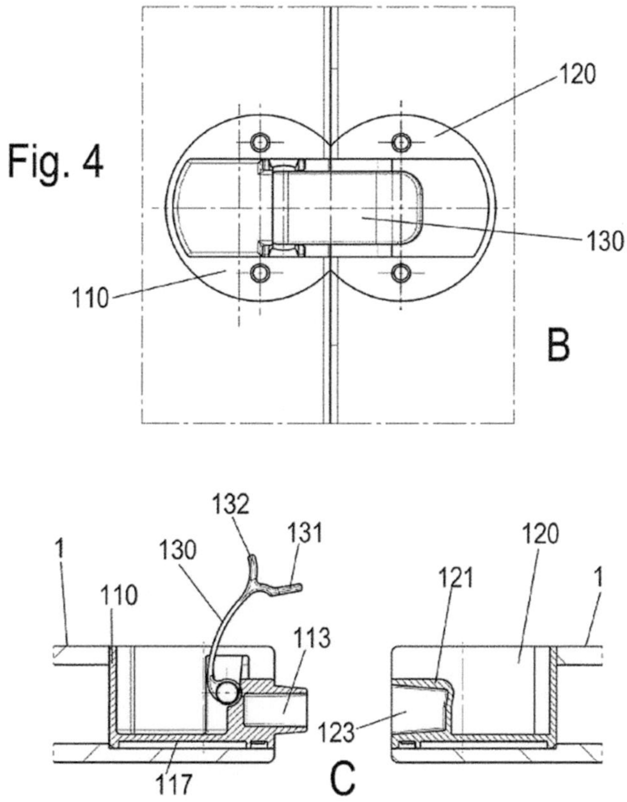

The document EP 3816412 A1 from the prior art discloses the connection of panels with a complementary stepped shape.[22]

Figure 8.23 shows the stepped connection form in a sectional view.

Figure 8.24 shows the fasteners of the prior art with the complementary structures of the panels in perspective views. With these fasteners, only straight connections can be made, not angled ones.

Patent specification WO2023099218 A1 discloses fasteners with a rotary lock.[23]

Figures 8.25 and 8.26 show the twist lock of the prior art, whose fastening elements 31 and 32 do not extend far into the respective panel, so that no mechanically stable connection can be achieved.

---

[22] DPMA, https://depatisnet.dpma/DepatisNet/depatisnet?action=pdf&docid=EP000003816412A1&xxxfull=1, accessed on 31.8.2024.

[23] DPMA, https://depatisnet.dpma/DepatisNet/depatisnet?action=pdf&docid=WO002023099218A1&xxxfull=1, accessed on 31.8.2024.

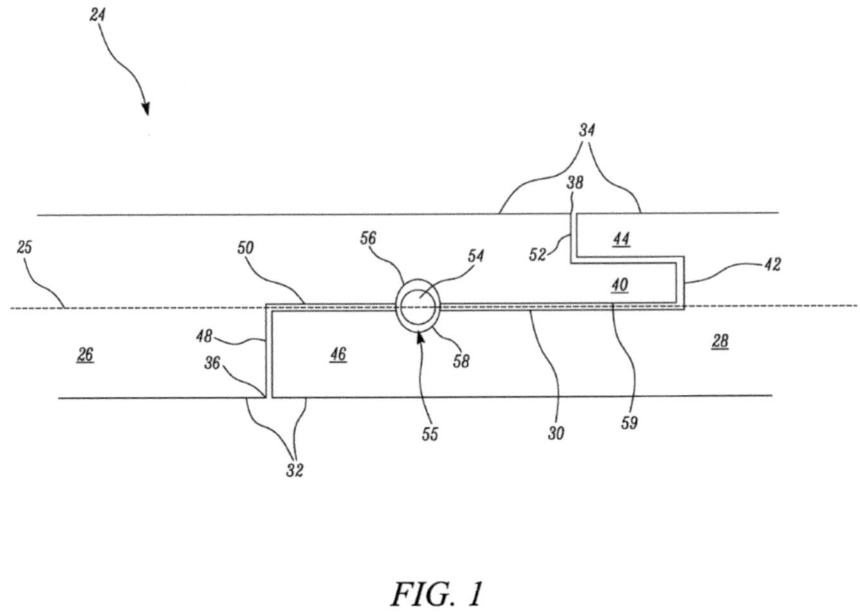

Fig. 8.23 Fig. 1 of EP3816412A1

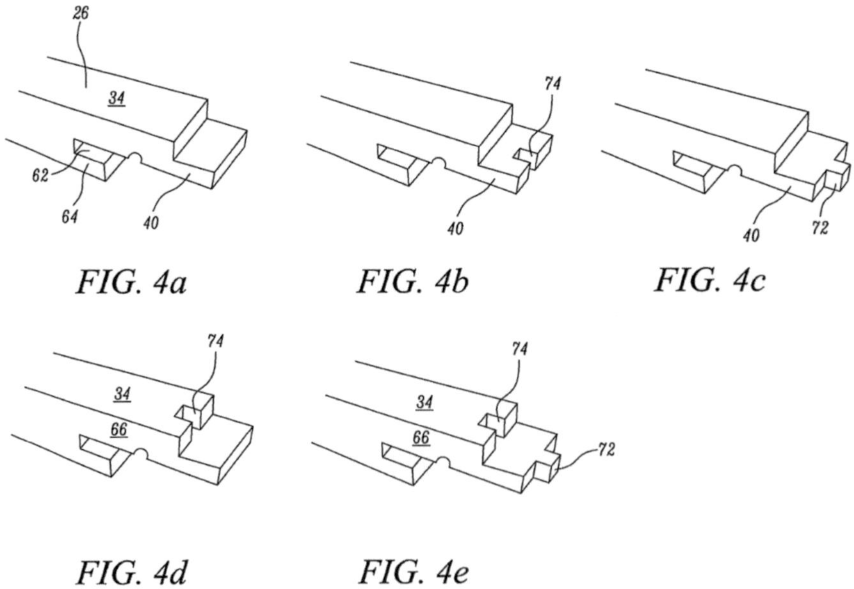

Fig. 8.24 Fig. 4a to 4e of EP3816412A1

Fig. 8.25 Figs. 1 and 2 of WO2023099218A1

Fig. 1

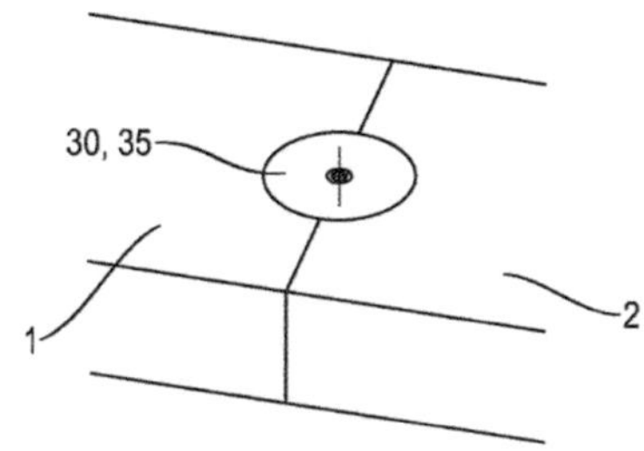

Fig. 2

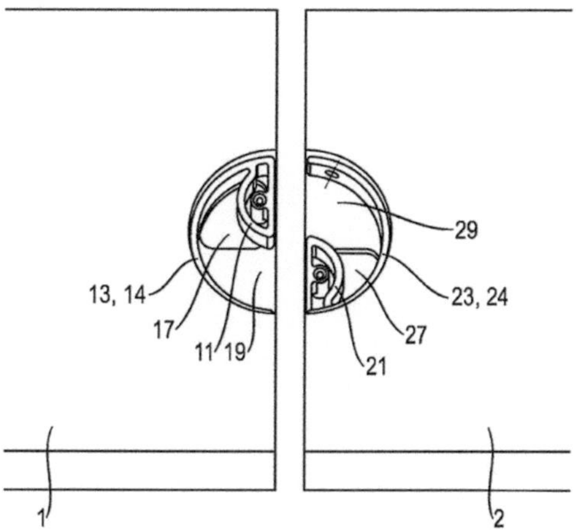

Fig. 8.26 Fig. 3 of WO2023099218A1

Fig. 3

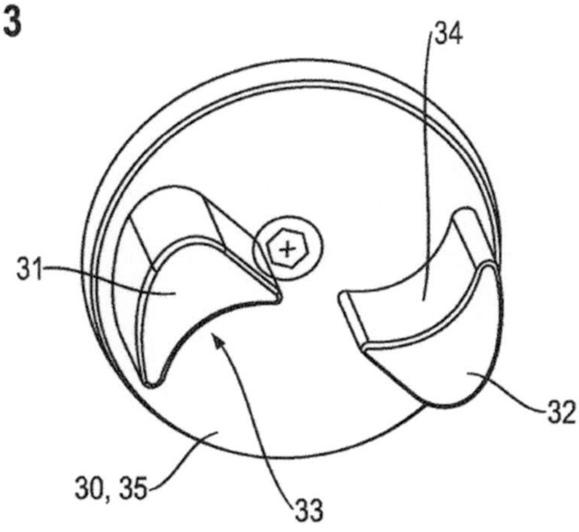

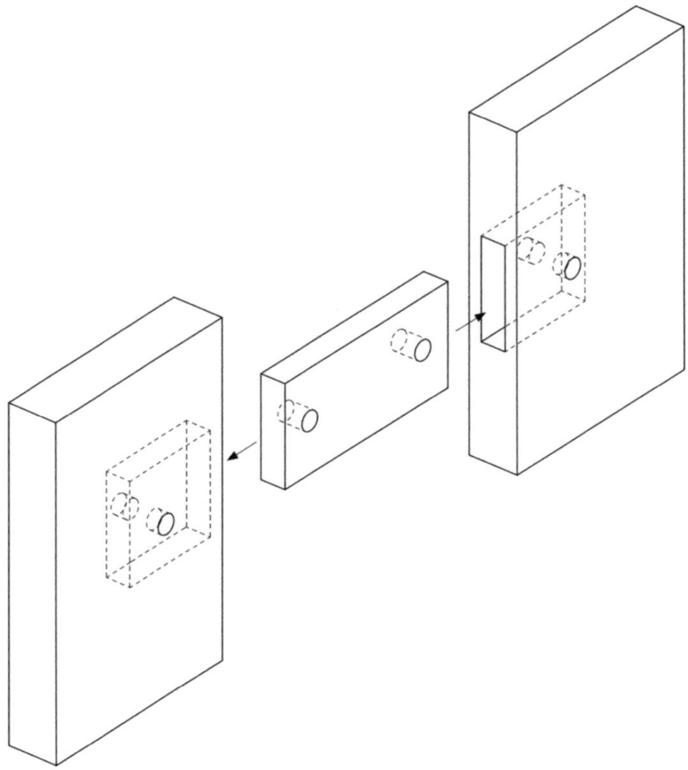

Fig. 8.27 Straight connector

Fig. 8.28 Right-angle connector

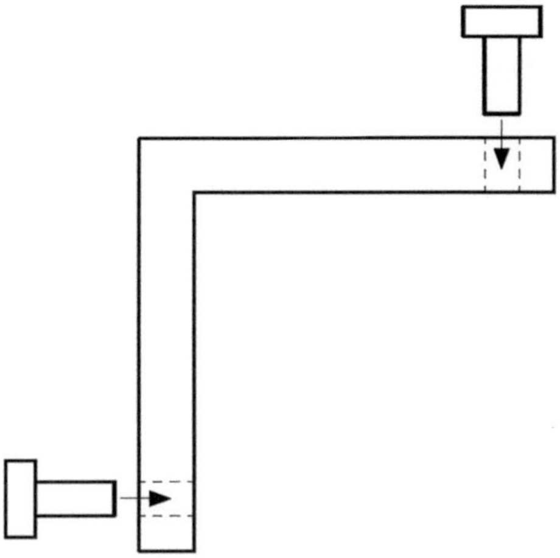

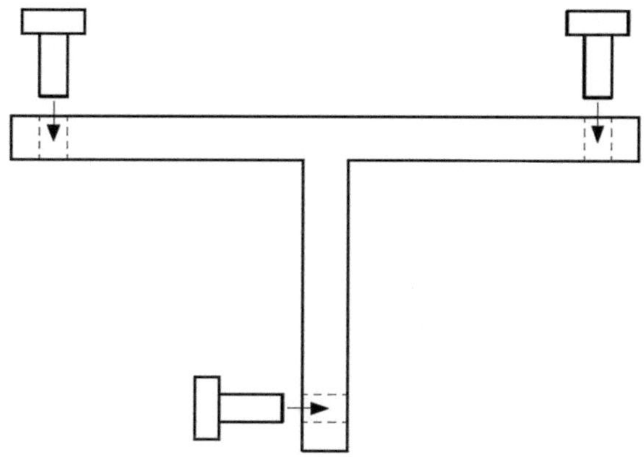

Fig. 8.29 T-beam connector

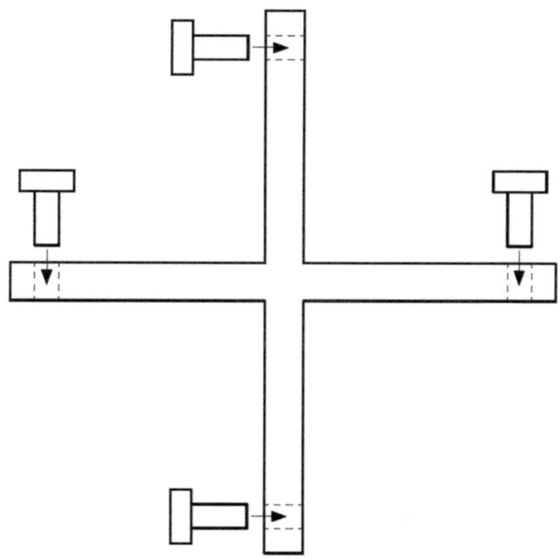

Fig. 8.30 X-shaped connector

Overall, the prior art thus presents the task of providing connectors for panels that enable a closed and mechanically stable connection without forming slots. In addition, not only flat but also angled connections should be made possible.

Figure 8.27 shows a proprietary connector that extends deep into the panels and can be screwed to the panels.

Figure 8.28 shows another proprietary connector for a right-angled connection of furniture panels.

Figure 8.29 shows a connector for right-angled joining of three panels, and Fig. 8.30 for joining four panels.

By means of connectors that extend deep into the respective piece of furniture and are shaped at right angles, or in T- or X-form, a proprietary technology can be provided for the production of modular, reconfigurable furniture.

MIX
Papier aus verantwortungsvollen Quellen
Paper from responsible sources
FSC® C105338

If you have any concerns about our products,
you can contact us on
ProductSafety@springernature.com

In case Publisher is established outside the EU,
the EU authorized representative is:
Springer Nature Customer Service Center GmbH
Europaplatz 3, 69115 Heidelberg, Germany

Printed by Libri Plureos GmbH
in Hamburg, Germany